Mines of *Goldfield, Bullfrog and other Southern Nevada Districts*

by F. L. Ransome

with Notes on the

Manhattan District

by Garrey and Emmons

This is a photographic reproduction of the 1907 U.S. Geological Survey Bulletin 303 entitled "Preliminary Account of Goldfield, Bullfrog, and Other Mining Districts in Southern Nevada."

Coverage includes Clark, Esmeralda and Nye Counties

Published by Stanley Paher

Nevada Publications
Box 15444
Las Vegas, Nev. 89114

ISBN 0-913814-60-1
Printed in the U.S.A.

CONTENTS.

Reprints from Out West Magazine 1907

ILLUSTRATIONS.

PRELIMINARY ACCOUNT OF GOLDFIELD, BULLFROG, AND OTHER MINING DISTRICTS IN SOUTHERN NEVADA.

By FREDERICK LESLIE RANSOME.

INTRODUCTION.

The following report aims to give in brief form such advance results as are available from a detailed investigation of the geology of the Goldfield and Bullfrog districts carried out during the autumn and winter of 1905–6. Complete reports on these districts are now in preparation and will be published as soon as possible. In the present bulletin are included also such reconnaissance notes as were made on outlying districts visited in connection with the principal work of the field season. Messrs. W. H. Emmons and G. H. Garrey, who had efficiently assisted in the work throughout the season, were instructed at its close to visit the Manhattan district, mainly for the purpose of ascertaining the area that should be included in future mapping and study of this new gold field. Their notes on the geology of the district are published as a separate section in this report.

During the summer and autumn of 1905 a topographic map, covering 8,500 square miles of southern Nevada and including the Goldfield and Bullfrog districts, was made by Messrs. R. H. Chapman and B. D. Stewart, topographers. A copy of this map accompanies this report (Pl. I). Mr. S. H. Ball was attached to Mr. Chapman's party as geologist, and the results of his reconnaissance of the whole region, made under the writer's general supervision, will be published as a separate bulletin (No. 308).

GOLDFIELD DISTRICT.

INTRODUCTION.

In June, 1903, Mr. J. E. Spurr paid a hasty visit to Goldfield, then known as the Grandpa district. Some indications of ore were known at this time, but no important ore bodies had been found. Mr. Spurr published a brief note[a] on the geology of the district, which was supplemented by a fuller statement,[b] made after a second visit in November, 1904.

During the summer of 1905 a topographic map on a scale of 2,000 feet to the inch and covering approximately 31½ square miles of the district was made by Mr. William Stranahan, topographer, and, with this map as a base, detailed geological work was begun in September of the same year by the writer, assisted by Messrs. W. H. Emmons and G. H. Garrey. Field work was completed in December.

The following account has been prepared to meet the demand for prompt information regarding a new and important district. It is not intended as a complete or final report on the geology and ores of Goldfield, but merely as a provisional advance statement, which will be amplified and modified when all the material gathered during the field work shall have been fully studied. Experience has shown that preliminary accounts of new mining districts are usually of value, but the benefit to the mining community may be gained at some cost to an author's professional reputation unless the object and limitations of such a preliminary outline are clearly understood by the reader. Therefore in this case it should be understood that the right to revise statements and conclusions is distinctly reserved, and when the final report appears it should entirely supersede the earlier publication.

HISTORY.

The discovery of Tonopah in 1900 marked, for Nevada, the beginning of a new era in its mining industry, which in the years following the decline of the Comstock had sunk to a very low ebb. As the size and richness of the Tonopah bonanzas became known new stimulus was given to prospecting and the desert ranges of southwestern Nevada were more thoroughly searched than ever before. Among the dozens of new camps that have come into existence within the last few years as a result of this activity, Goldfield, situated in Esmeralda County, 24 miles nearly due south of Tonopah (see fig. 1 and Pl. I, in pocket), has attained chief importance.

"Float" gold was found late in 1902 by Harry Stimler and William Marsh on Columbia Mountain and, after seeking in vain for the source of this gold, these early prospectors worked northward and on December 2 located the Sandstorm lode. A brief period of excitement followed the discovery of gold, but, as no important ore bodies were immediately uncovered, interest flagged and only a few determined men remained in the district. Among these were A. D. Myers and R. C. Hart, who, on May 24, 1903, located the Combination lode

[a] Bull. U. S. Geol. Survey No. 225, 1904, pp. 118–119.
[b] Bull. U. S. Geol. Survey No. 260, 1905, pp. 132–139.

and began at once to develop their property. In October of the same year, after the existence of ore had been proved by a short tunnel, the ground was bonded to the Combination Mines Company and in the following December this company began the shipment of ore.

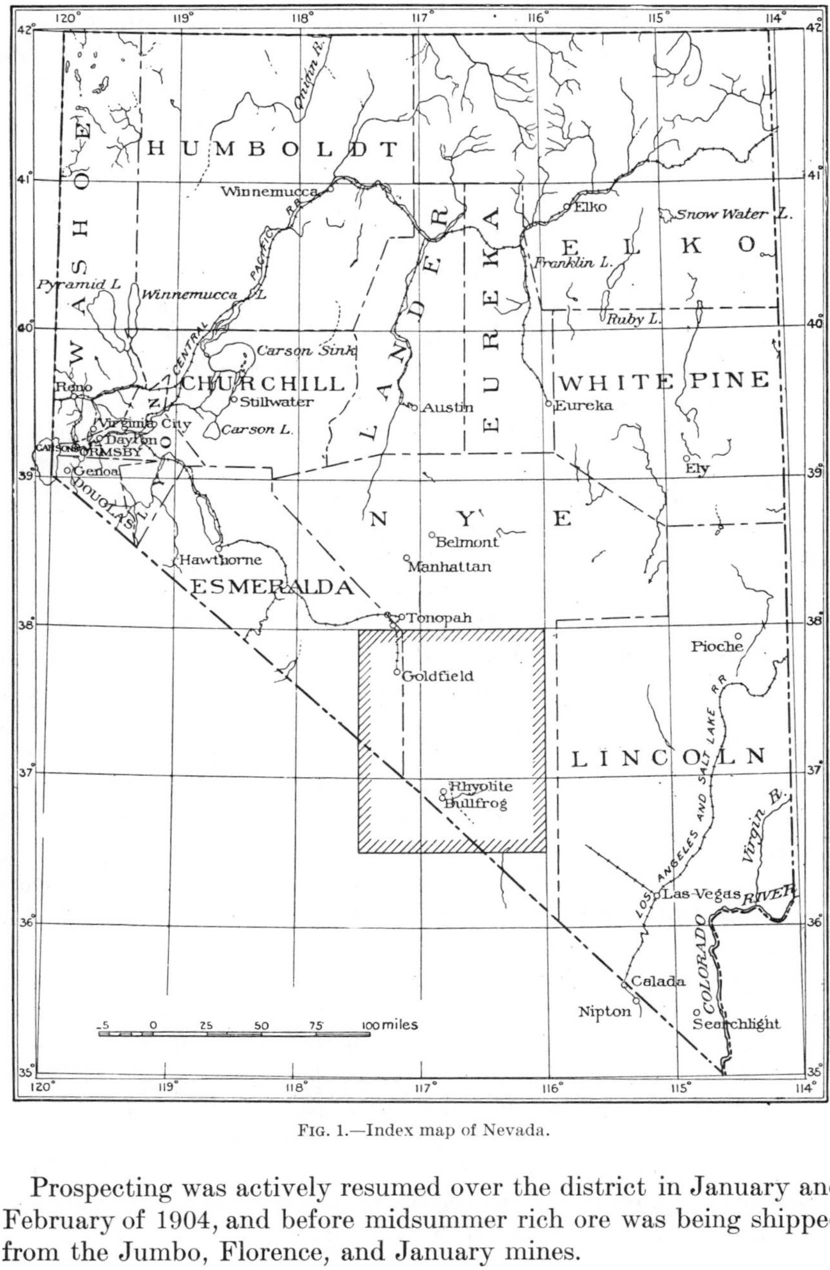

FIG. 1.—Index map of Nevada.

Prospecting was actively resumed over the district in January and February of 1904, and before midsummer rich ore was being shipped from the Jumbo, Florence, and January mines.

With the exception of the Combination, all the mines which have attained any importance were first opened and developed by the leasing system, and so near the surface and so rich were many of the ore bodies that numbers of the successful lessees made ample fortunes in a few months or even in a few weeks. The royalties usually exacted ranged from 20 to 30 per cent of the net value of the ore extracted. At one time there were six sets of lessees working through as many shafts on the Jumbo lode. The blocks were from 100 to 300 feet in length, the most productive and best known being that generally referred to as the Bowes-Kernick lease.

The daily output from all of these workings in August, 1904, was estimated at $10,000, and the total to the expiration of the leases, early in 1905, amounted to considerably more than $1,000,000. Another famous lease was the Sweeny lease, on the Florence mine, which expired on January 15, 1905, after a production amounting to $650,000. The Patrick lease on the January mine, probably the first important lease granted in the camp, produced about $500,000. Later, and in some respects more sensational than the leases just mentioned, was that granted to B. J. Reilly and associates on No. 1 block of the Florence ground. This lease, after having been successively abandoned by two earlier holders, was taken up by Reilly and his partners in April, 1905. Ore was found in the following June, but it was not until August that the true richness of the body became apparent. In the short time then available before the expiration of the term, on October 19, 1905, nearly all of the ore which the lease produced was mined and hoisted to the surface. The royalties paid to the Florence-Goldfield Mining Company represent a total output of $475,000. Most accounts, however, give the production as $750,000. Some of this ore was of extraordinary richness, a lot of about 4½ tons, taken out from the bottom of the workings during the last days of the lease, averaging, according to Mr. Frank Oliver, superintendent of the Florence mine, between $6,000 and $7,000 per ton.

Early in 1905 the district probably contained fully 8,000 people, over two-thirds of whom were in the town of Goldfield. Late in the year, owing to the attractions of Manhattan and other recently opened camps and to a decreasing activity in prospecting, in leasing, and in the various more or less speculative transactions so characteristic of a new mining district, the population had somewhat declined.

The completion of a railroad from Tonopah to Goldfield in September, 1905, gave the district a much-needed improvement in transportational facilities, although freight charges are still maintained at unduly high figures. Another important event of the year was the discovery of a large body of rich ore in the Redtop mine, which had not previously been a productive property.

PRODUCTION.

In 1904 the Goldfield district produced over $2,300,000. Statistics for the year 1905 are not yet available, but it is doubtful whether the output for the second year of activity will exceed $2,500,000.

TOPOGRAPHY.

The hills in the vicinity of Goldfield form an irregular group (see Pl. I), having no definite trend and belonging to no recognized mountain range. On the north, east, and south these hills slope down to broad undrained sterile basins, each with its characteristic central playa—a gleaming level plain incrusted with salts, except when an occasional shower has spread a film of water over its surface. The average altitude of the playa floors above sea level in this region is about 5,000 feet. Such a basin, approximately 10 miles in width, separates the Goldfield Hills from Stonewall Mountain, a prominent landmark southeast of the district. A low, broad pass between this basin and one to the north of it connects the Goldfield Hills with the northwest end of the Cactus Range, Cactus Peak being 20 miles east-northeast from Goldfield. A similar inconspicuous divide, about 12 miles north-northeast of Goldfield, forms a slight connection between the Goldfield Hills and the hills of the Southern Klondike and Tonopah districts.

West of town a basaltic mesa, at this latitude about 4 miles wide, but becoming much broader toward the south, stretches to the eastern base of Montezuma Peak, which is composed in part of rocks similar to those of the Goldfield district. The old mining settlement of Montezuma lies near the summit of the peak, which is 8,426 feet in height and 7 miles a little south of west from the town of Goldfield.

Goldfield is situated on the southern rim of one of the typical desert basins of the region, which merges through a low pass 22 miles north of town into the still larger basin west of Tonopah. This pass, through which the railroad runs from Tonopah to Goldfield, is between the Lone Mountain Range on the west and the Tonopah Hills on the east.

The portion of the Goldfield Hills which is particularly described in the present paper and which will for convenience be hereafter referred to as the Goldfield district is shown in the accompanying map (Pl. II), a much reduced and simplified copy of the geological map that will be published with the final report. The relief in this area ranges from 5,400 feet in its northwest corner to 6,635 feet on the summit of Preble Mountain, near the southeast corner. The culminating peak of the Goldfield Hills, however, is just east of the area represented in the map and attains an elevation of nearly 7,000 feet. A large part of the district may be described as a rolling plateau (see

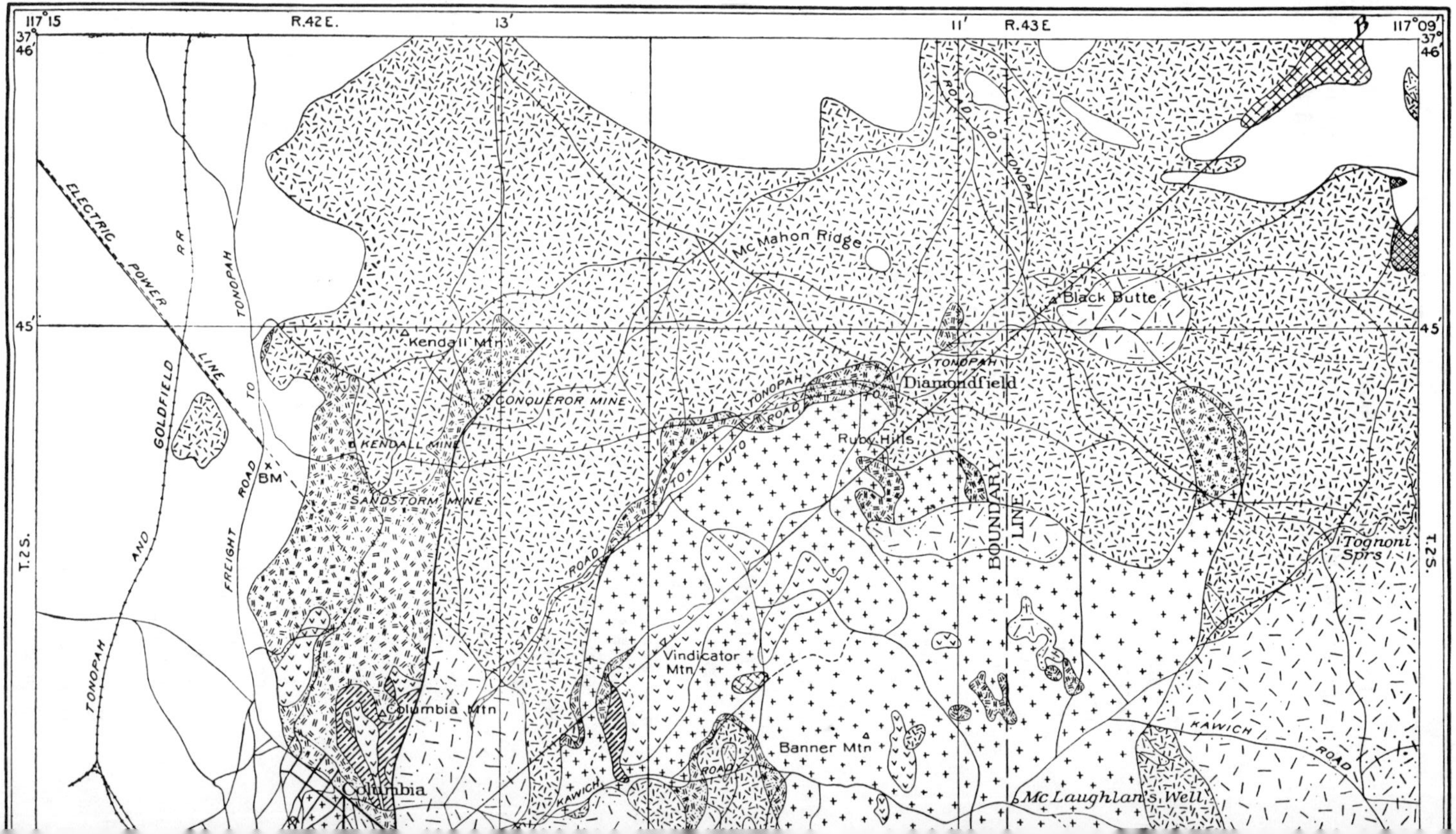
117°15'
37°46'
R.42 E.
13'
11'
R.43 E.
117°09'
37°46'
45'
T.2 S.
ELECTRIC POWER LINE
GOLDFIELD AND TONOPAH RR
TONOPAH TO GOLDFIELD FREIGHT ROAD
BM
Kendall Mtn
CONQUEROR MINE
KENDALL MINE
SANDSTORM MINE
Mc Mahon Ridge
ROAD TO TONOPAH
Black Butte
Diamondfield
TONOPAH TO AUTO ROAD
STAGE ROAD
Ruby Hills
BOUNDARY LINE
Tognoni Sprs
Vindicator Mtn
Columbia Mtn
Banner Mtn
Columbia
KAWICH ROAD
Mc Laughlan's Well

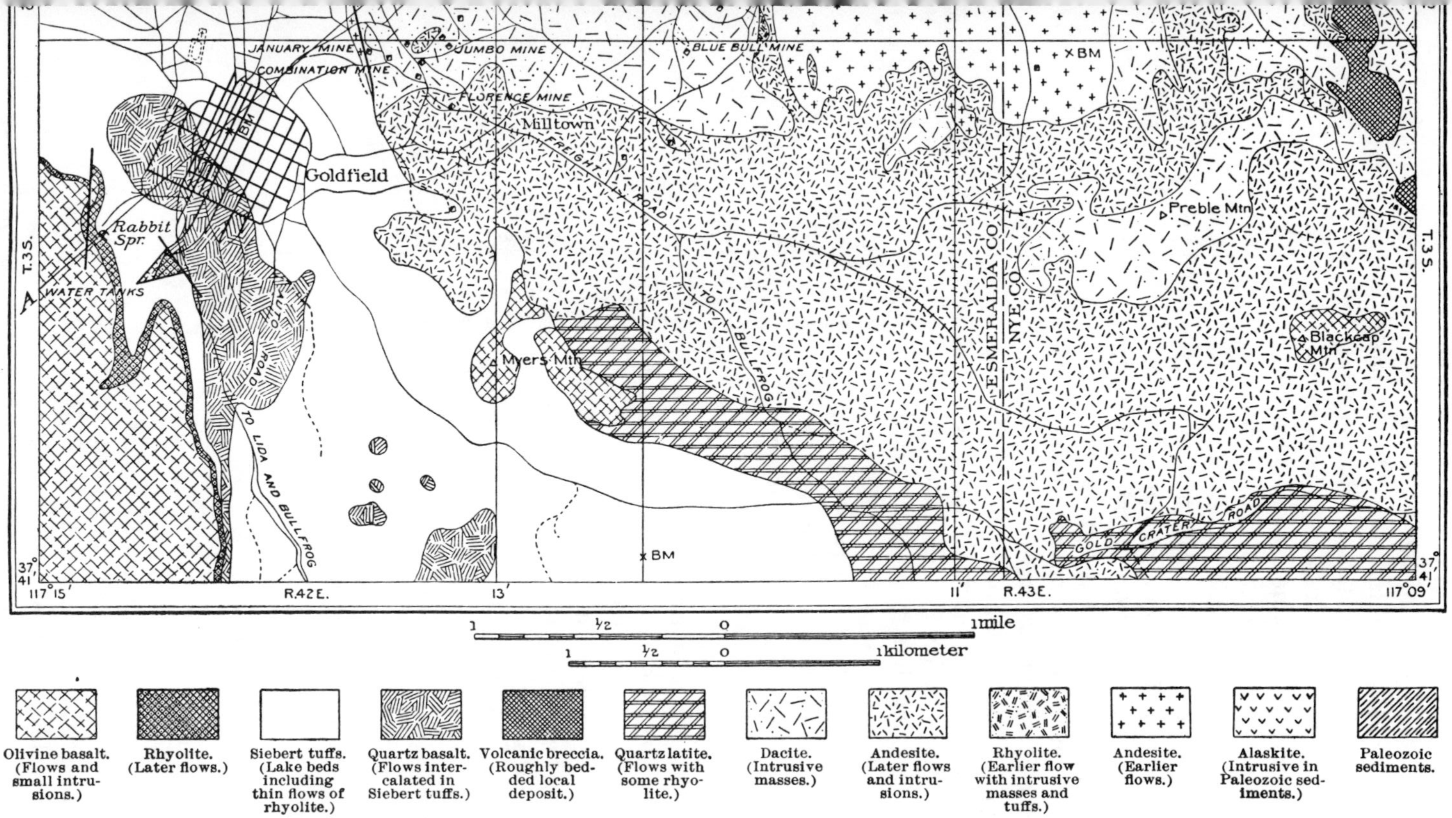

PRELIMINARY GEOLOGIC MAP OF THE GOLDFIELD DISTRICT.

fig. 2), ranging from 5,700 to 5,800 feet in general elevation and surmounted by many irregularly distributed peaks and knobs, of which Preble, Blackcap, Vindicator, Banner, and Columbia mountains are conspicuous examples. None of these hills is a volcanic crater, as has been asserted in some descriptions of the district, nor are they due to local volcanic upheaval. They owe their height above the surrounding country merely to the fact that they are composed of material somewhat harder or more durable than the average rock of the region and have thus succumbed more slowly to the general erosion that has carved the existing relief from a gently domed volcanic series. The superior resistance of the hills is in some cases due to the original character of the rock of which they are composed; in other cases, to a capping of young and fresh lava, but by far the greater number of the smaller knobs that stud the district, of which Black Butte is an excellent example, are prominent because the rock has been hardened by local silicification in a manner that will be described in connection with the ore deposits.

The fact that much of the district is a rolling plain diversified by rocky knobs has an important bearing in a region where detritus tends to accumulate on all but the steepest slopes. In the Tonopah district, and especially in the Bullfrog district, there is generally a fairly definite line between steep, bare, rocky hills and the gentle alluvial slopes at their base. At Goldfield, however, the general gradients, as may be seen from fig. 2, are so slight that detritus accumulates over much of the surface and interferes seriously with attempts to determine the distribution and structure of the underlying rocks.

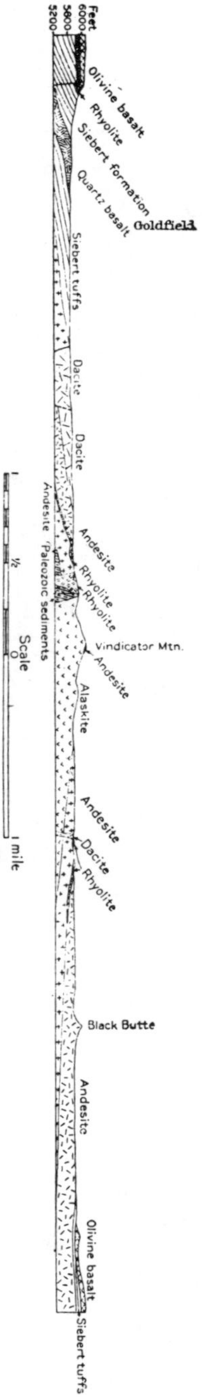

FIG. 2.—Geologic section across the Goldfield district on line *A–B* of Pl. II.

GENERAL GEOLOGY.

INTRODUCTION.

The geological structure of Goldfield compared with that of most districts in which deposits of the precious metals are found is simple. The district is essentially a low, domical uplift of Tertiary lavas and tuffs resting upon a foundation of ancient granitic and metamorphic rocks. The erosion of this flat dome has exposed the pre-Tertiary rocks at a number of places in the central part of the district, and these outcrops are surrounded by wide concentric zones of successively younger formations. But while it is not difficult to recognize the broad, bare outlines of the structure, this general idea requires modification in many particulars before it can be accepted as a satisfactory conception of the geology of the region, and it is in the study of these essential details rather than in complexity of structure that the investigation of the district offers real difficulties. One of these obstacles, the extensive cloaking of the surface by gravelly alluvium or "wash," has already been referred to. Owing to the prevalence of this material the contacts between the various underlying rocks are seldom well exposed and are usually completely covered. While it would be a simple matter to map this detritus wherever it is thick enough to conceal the rock beneath it, the map thus made would be less useful to the miner than one in which an attempt has been made to show the distribution of the possibly ore-bearing rocks, observed facts being supplemented wherever necessary by inferences based upon all available data. In accordance with this general principle the areas of alluvium have been entirely omitted from the preliminary map forming Pl. II and will be reduced to a minimum in the complete and final map. That some adjustment of geological boundaries will be demanded when mining operations determine the positions of contacts now concealed is of course to be expected; but it is deemed more important to present a helpful interpretation of hidden structure than to be content merely with cautious and unsuggestive adherence to visible facts.

Another obstacle to accurate and successful study of the region is the intensity, widespread distribution, and erratic character of the alteration that has affected rocks, some of which, even in a fresh state, could be distinguished from one another only by refined petrographic methods.

PRE-TERTIARY ROCKS.

The oldest rocks exposed in the district are found on Columbia and Vindicator mountains, and are particularly noticeable on the southern slope of the former peak as conspicuous dark patches visible from Columbia or Goldfield. Originally thin-bedded limestones or calcareous shales, with subordinate beds of sandstone, the formation is now much contorted and metamorphosed. The calcareous beds have been changed to compact aggregates of quartz (jasperoid) crowded with microscopic black particles, probably carbonaceous, and the sandstone to hard, vitreous quartzite.

The metamorphism is apparently due to the intrusion of these sediments by masses of a granitic rock composed essentially of quartz and orthoclase, a type for which Mr. Spurr has proposed the name alaskite. This rock is well exposed on Columbia Mountain and makes up a large part of Vindicator Mountain. Smaller areas occur in the vicinity of Banner Mountain. Most of the alaskite contains a very little black mica and some plagioclase. In most exposures the mica is decomposed and scarcely noticeable, but northeast of the summit of Vindicator Mountain the alaskite mass grades into a biotite granite.

The age of the alaskite and intruded sediments, beyond the fact that both are pre-Tertiary, can not be determined from any evidence available in the Goldfield district. Mr. Spurr,[a] whose wide experience in southern Nevada renders his opinion of special weight, considers that the metamorphic sediments may well be Paleozoic and that the alaskite was intruded in post-Jurassic time. Mr. S. H. Ball,[b] who recently made a reconnaissance over a large area around the Goldfield district, found Cambrian fossils in beds which he is inclined to correlate with those exposed in Columbia and Vindicator mountains.

It is certain that the intrusion of the alaskite was followed by a long interval during which the region was eroded to the surface of moderate relief subsequently covered by the irruption of Tertiary lavas.

TERTIARY VOLCANIC ROCKS.

The sequence of Tertiary eruptions as shown by the rocks of the Goldfield district was as follows, the oldest rock being at the bottom of the column:

Olivine basalt; flow and small intrusive bodies.
Breccia and conglomerate.
Rhyolite; flow.
Lake beds.
Quartz-bearing basalt; flow intercalated in lake sediments.
Lake beds.
Latite and rhyolite; flows.
Dacite; sheetlike and irregular intrusions, possibly in part a flow.
Andesite; flows with some tuff and small intrusive bodies.
Rhyolite; intrusive masses.
Rhyolite; flows and tuffs.
Andesite; flow.

The older or first andesite, as may be seen from Pl. II, occupies an area a little over 2 miles in diameter in the center of the district, and a smaller area south of Columbia. Its characteristic features may be well seen in the vicinity of McLaughlan's well, on Banner Mountain, and on the southern edge of the town of Columbia. This rock has not been found in a fresh condition. Its color is usually dull purplish or greenish gray. A spotted or mottled appearance, due to flow-brecciation, is rather characteristic and is well shown on a small hill half a mile north of the summit of Banner Mountain. The freshest obtain-

---

[a] The ores of Goldfield, Nev.: Bull. U. S. Geol. Survey No. 260, 1905, p. 133.
[b] Oral communication.

able specimens of the first andesite show phenocrysts of feldspar, hornblende, and mica in a compact, streaky groundmass that was evidently once a glass with well-marked flowage lines. All the phenocrysts have the dullness that indicates more or less thorough alteration and the original glassy base is now devitrified. As a rule, the first andesite, when it has not been subjected to the intense local alteration connected with ore deposition, may be distinguished from the later andesitic rocks by the absence of fresh phenocrysts and the peculiar streakiness of its groundmass, which resembles the groundmass of a rhyolite rather than that of most andesites.

The rock is apparently a flow resting upon a clean and rather uneven surface of alaskite and Paleozoic sediments. Its thickness is not known, but is probably very variable, as on Columbia Mountain the succeeding rhyolite rests directly upon the alaskite. The maximum thickness is tentatively estimated as at least 400 feet, and it may be considerably more.

Overlying the first andesite, or resting on the pre-Tertiary rocks where the andesite is absent, is a series of thin glassy rhyolite flows which is best developed in the rounded white hills northwest of Columbia Mountain. This rhyolite is associated with pale-greenish tuffs of mingled andesitic and rhyolitic character and is cut by masses of intrusive rhyolite. All these rocks are shown together in Pl. II, although they will be distinguished on the final map. The effusive rhyolite is usually nearly white, shows conspicuous flow lamination, and weathers in thin, shelly flakes. This is the rock of the Sandstorm mine and of the hills lying west of the road from the mine to Columbia. It all shows some devitrification and silicification, particularly in the vicinity of the ore deposits. Certain varieties or portions of this rhyolite acquired by flowage a very thin lamination, causing the rock when somewhat weathered to closely resemble indurated clay shale. This appearance is very deceptive, as may be seen at the Tonopah Club or Conqueror mine. The town of Diamondfield is partly underlain by this shalelike rhyolite, and small residual patches of similar material are scattered over the surface of the first andesite. The rhyolite appears to have been thickest in the vicinity of Columbia Mountain and to thin out to the southeast. None of it has been found southeast of McLaughlan's well.

The intrusive rhyolite occurs on the summit of Columbia Mountain, where it has locally a sheetlike form and constitutes the ridge extending north of the mountain to the Adams mine. It is intrusive into the shaly rhyolite at the Conqueror mine, and another small mass cuts the first andesite, alaskite, and pre-Tertiary block jasperoid on the west slope of Vindicator Mountain. Thin sections indicate that some of this intrusive rhyolite occurs also on the east slope of Vindicator Mountain, but owing to the intensity of the alteration in this vicinity it could not be separated in the field from the silicified first andesite. The tuffs associated with these early rhyolitic eruptions may be best examined along the southwest base of Columbia Mountain and directly west of the Kendall mine. Underground they are exposed in the west workings of the 100-foot level of this mine.

After the eruptions of the earlier rhyolites came the eruptions of the second andesitic period, which covered the region with successive flows of andesite, accompanied by subordinate layers of andesitic tuff. The products of these eruptions constitute the most extensive formation in the district. (See Pl. II.) The total thickness of this deposit, as of most of the other volcanic formations in this district, can not be determined, owing to the slight deformation of the rocks, their poor exposures, the comparatively slight topographic relief, and the absence of persistent datum planes, such as may usually be found in stratified rocks. It probably lies between 600 and 1,000 feet; but this is confessedly little more than a guess and is likely to be under rather than above the truth.

The petrographic study of the andesites has not been finished. They contain in general phenocrysts of labradorite feldspar, hornblende, augite, orthorhombic pyroxene, and occasionally biotite in groundmasses ranging from hyalopilitic to vitrophyric in texture. The hornblende exhibits various stages of magmatic resorption, the abundance of the augite being in rough inverse proportion to the diminution of the hornblende by this process. The second andesite undoubtedly includes more than one flow, but it is quite impracticable in this field to discriminate individual flows in mapping the whole area. Preliminary examination of over 200 specimens and thin sections of the andesites indicates that any division based on the predominance of certain ferromagnesian constituents—for example, augite andesite or hornblende andesite—would be artificial. In other words, the relative proportions of these minerals appear to vary in any one flow. Moreover, as the first andesite and pre-Tertiary rocks are cut by dikes and small intrusive bodies of the second andesite, the latter formation as a whole is probably a volcanic complex, in which are involved intrusive bodies, as well as flows and beds of tuff and breccia.

The dacite, whose eruption followed the second andesitic period, is in many respects the most interesting rock in the district and is certainly the most important from the economic standpoint. It occupies, as may be seen by reference to Pl. II, a considerable area east of Goldfield and Columbia and is the principal country rock of the Combination, January, Jumbo, Redtop, and Florence mines. The surface at the main Florence shaft is the second andesite, and it was at first supposed that the ore also occurred in this rock. Microscopic study shows, however, that most of the altered country rock of the Florence below the zone of complete oxidation, although finer grained than the typical dacite, is really a variety of that rock, and consequently that the shaft penetrates the flat-lying andesite and enters the dacite. This mass of dacite appears to be an intrusive sheet dipping gently to the west and south and cut off to the west by faulting. Its structural relations will be more fully described in subsequent pages.

A larger body of dacite extends southward from Tognoni Springs and forms some of the high hills just east of the mapped portion of the district. This mass, as may be seen from Pl. II, makes up a large part of Preble Mountain. Smaller bodies of dacite occur southeast of Black Butte and in the central part of the district.

The dacite, of which typical and fairly fresh specimens can be collected on the little hill 500 feet nortneast of the Transformer station, has the appearance of an andesite. It is generally more conspicuously porphyritic than the andesites of the district, but its essential distinction from them is the presence of phenocrysts of quartz. These are always rounded by magmatic corrosion and as a rule are neither very numerous nor very conspicuous. The rock consists of labradorite feldspar, hornblende, biotite, augite, and quartz in the usual andesitic groundmass. Chemical analysis of the dacite shows that it contains 60 per cent of silica, or only 2 per cent more than a hornblendic andesite from the northeastern part of the district. Mineralogically and chemically the dacite at Goldfield is merely a quartz-bearing andesite and is in no way similar to the more siliceous rocks called dacite by Mr. Spurr in his report on the Tonopah district.[a]

The geological relation of the dacite to the andesite is not entirely clear. Its comparatively coarse porphyritic texture, the irregular outlines of its exposures (see Pl. II), and the occurrence of small bodies of dacite within larger masses of andesite are all indicative of the intrusion of the dacite into the andesite. As a rule, however, the contact between the two rocks is not well enough exposed or the rocks have been too much altered in its vicinity to afford definite proof of this relation. At one point only, situated 1 mile N. 35° E. of the summit of Preble Mountain, a sharp intrusive contact is exposed between the dacite and the andesite. The dacite is glassy at the contact and penetrates the andesite in a way that shows the latter to be the older rock. Of course this exposure merely proves that one mass of the dacite is intrusive into a certain mass of andesite. The general conclusion reached, that the dacite as a whole closely followed the andesitic eruptions and was intruded into the flows and subordinate tuffs at moderate depth, is still in need of evidence for its complete establishment. It is probable that this intrusion was accompanied

[a] Mr. Spurr's reasons for using the name dacite for rocks which should strictly receive other names are given in Geology of the Tonopah mining district, Nevada: Prof. Paper U. S. Geol. Survey No. 42, 1905, p. 59, footnote.

by some deformation of the region, and the andesite flows appear to have been somewhat eroded before the next eruptions took place.

After the dacitic intrusion a rock provisionally classed as quartz latite was erupted as a series of flows and is now extensively exposed in the low hills in the southeast corner of the district. This rock, which corresponds more closely than anything else in the Goldfield region to some of the Tonopah dacite, is in most places a porous light-gray or greenish rock with abundant glistening scales of black mica and phenocrysts of fresh plagioclase, mostly oligoclase. In some varieties quartz and orthoclase are present, and some thin flows of vitrophyric rhyolite are associated with the quartz latite. Two areas of rhyolite in the northeast corner of the district are provisionally mapped with the latite in Pl. II. The quartz latite is easily worked, and a quarry opened in this material 1½ miles south of Preble Mountain has afforded stone for buildings in Goldfield.

No traces of mineralization have been found in the latite, and the ore deposits of the district were probably formed prior to its eruption. This, however, is not definitely determinable, as none of the latite occurs in the immediate vicinity of known ore bodies.

After the latitic eruptions the region was covered by an extensive body of fresh water in which were laid down 500 feet or more of bedded conglomerates, tuffs, tuffaceous sands, pumiceous ash, and diatomaceous earth. These deposits, the "cement rock" of the miners, cover a large part of the district, particularly in its outskirts. They underlie the town of Goldfield and stretch both northward and southward beyond the limits of the area studied. They also extend beneath the "malapi" (malpais) mesa southwest of town and are well exposed in the steep bluffs of the ravine in which are situated the city waterworks. The beds here dip generally westward at angles ranging from 25° to 30°.

Although the lake beds appear to be locally conformable with the quartz-latite flows, yet as these are present beneath them in only a small part of the district, and as the formation of the lake indicates some change in the configuration of the region, there is at least a suggestion of erosion between the quartz-latite eruptions and the lacustrine condition. Volcanism continued active during the existence of the lake, as shown by the prevailingly volcanic nature of the sediments and the occurrence within them of a thick flow of quartz-bearing basalt and thin flows of glassy rhyolite. The quartz basalt, as shown by wells, underlies the southwestern part of Goldfield and is exposed on the automobile road to Bullfrog.

In the eastern part of the district, northeast of Preble Mountain, are some bedded breccias and tuffs from 200 to 300 feet in present thickness. They rest upon dacite and andesite and are made up largely of fragments of these rocks, particularly of dacite mingled with abundant fragments of volcanic glass and some sandy tuffs. These beds are provisionally correlated with the lake beds just described.

Lacustrine conditions were ended by deformation of the region, the domical structure being accentuated or possibly initiated by this uplift. After the lake beds had been beveled by erosion and probably entirely swept away from the central part of the district eruptions of rhyolitic lavas again began.

The latest rhyolite is a light-pink or pale-brown rock, rather porous and containing abundant pumiceous fragments. It forms a flow about 100 feet in maximum thickness, which is exposed in the bluffs southwest of Goldfield, under the dark volcanic rock which caps the mesa. It is this rhyolite that is quarried for building stone at Rabbit Spring. It is in some places overlain by a few feet of unconsolidated tuffaceous sand and volcanic gravel.

Finally, the volcanic history of the region was brought to a close by the eruption of one or more flows of olivine basalt. This is the dark rock that is conspicuous as the capping of the mesa southwest of town. It also forms the summit of Myers Mountain, Blackcap Mountain, and several hills south and east of the area mapped. The larger areas of this rock represent remnants of flows which rest upon a nearly horizontal surface of older rocks; but some smaller masses, such as that shown on the east slope of Vindicator Mountain, are intrusive and may fill orifices through which the basalt reached the surface to spread out as flows.

CORRELATION.

While final correlation of the Goldfield rocks with those of Tonopah and other districts can not be attempted in this preliminary report, yet it may be suggested that the lake beds of the Goldfield district probably correspond to the Siebert tuff[a] of the Tonopah district, which Mr. Spurr regards as probably of Miocene age. If this correlation proves to be correct, the Tonopah rhyolite dacite of Mr. Spurr will correspond to the quartz latite of the Goldfield sequence. The dacite of the Goldfield district has apparently no representation at Tonopah and it does not seem possible at present to correlate the two andesitic formations at Goldfield with the "earlier andesite" and "later andesite" at Tonopah. The oldest andesite at Goldfield is petrographically unlike either of the Tonopah andesites and was succeeded by rhyolites having no representation in the northern district.

Olivine basalt occurs at one point in the Tonopah district, and as it overlies the tilted Siebert tuff it is probably of the same age as the olivine basalt of the Goldfield district. The intrusions of rhyolite represented by Mount Oddie and of "dacite" represented by Butler,

[a] Spurr, J. E., Geology of the Tonopah mining district, Nevada: Prof. Paper U. S. Geol. Survey No. 42, 1905, pp. 51-55.

Brougher, and Siebert mountains, as well as the "latest rhyolite-dacite" flow on the slopes of Oddie and Brougher mountains, appear to be later than any of the volcanic rocks in the Goldfield district.

GEOLOGICAL STRUCTURE.

It has already been said that the district has the general structure of a low dome, the younger formations dipping gently away from a nucleus of slightly upheaved pre-Tertiary rocks that have been exposed here and there by erosion. The concentric regularity of outcrop of the various formations about the structural center of the district, which may be considered to lie in the vicinity of Banner Mountain, is modified by two main factors. One of these is the intruded dacite; the other is a strong fault which extends from a point east of Kendall Mountain southward along the east side of Columbia Mountain to the vicinity of the Redtop mine. Here it is apparently offset by a cross fault (Pl. II), but it continues southward beyond this offset past the January mine until it disappears under the lake beds and alluvium east of Goldfield. This dislocation may be conveniently referred to as the Columbia Mountain fault.

The only body of the dacite whose structural relations can here be considered in any detail is that lying just east of the towns of Goldfield and Columbia and containing the Combination, January, Redtop, and Jumbo mines. The general relations of the boundary of this mass to the topography indicate that it is a rather irregular sheet with a gentle southwest dip. It is supposed to be intrusive, but present exposures of this particular mass afford no conclusive evidence to support this view as against the idea that it is a flow. This dacite, as indicated by the character of its southern contact and as shown by the workings of the Florence mine, extends southward under the andesite between Goldfield and Milltown. The small area of andesite shown in Pl. II between the Combination and Jumbo mines is probably an outlying remnant of the same andesitic flow that laps up over the dacite at the Florence mine.

The thickness of this dacite sheet is not known. The Jumbo shaft, sunk at a point where apparently very little of the dacite has been eroded away, was 450 feet deep at the end of 1905 and was still in this rock. Such data as are available indicate that the dacite sheet has an average dip rather under than over 10°. If the sheet was of uniform thickness prior to erosion, then the bottom of the Jumbo shaft should be very near the bottom of the dacite. It is to be hoped that sinking at this point will be continued, as the determination of the thickness of the dacite at this particular place will be of much value in forecasting the probable changes in country rock that will be experienced in the other mines with increased depth. Regularity in an intruded sheet can not, however, be assumed, and the dacite may be much thicker in the vicinity of the principal mines east of Goldfield than is indicated by its general relation to the topography. The preparation and study of several geological structure sections, such as are planned for the final report, will probably throw additional light on the form and thickness of the dacite mass.

The Columbia Mountain fault dips to the east and has a normal throw, dacite and andesite being dropped against the older rhyolite and pre-Tertiary rocks of Columbia Mountain, as shown in Pl. II. The fault is very poorly exposed. It is followed by an inclined shaft at the Conqueror mine and is cut by the Columbia tunnel on the east slope of Columbia Mountain. Between these two points the outcrop of the fault is covered by loose material, and there are no mine workings to reveal the character of the fissure. On the south slope of Columbia Mountain the presence of the fault determines the position of a little ravine eroded in the soft, crushed material of the fault zone. The distribution of the rocks south of Columbia Mountain shows that the main fault is offset by a transverse fault having approximately the position shown in Pl. II. The rocks in this vicinity, however, are much altered and are so covered by detritus that no exposures of this fault are known. It apparently passes just southwest of the Redtop mine into an area where future mining development will probably afford satisfactory proof of its presence. South of Columbia the position of the main fault is only approximately determinable as a general line of demarcation between the locally much-altered first andesite on the west and the dacite on the east. The fault must pass only a short distance west of the January shaft, but it is not exposed at the surface. At the bottom of the January mine, however, 280 feet below the surface, a very soft zone of crushed and altered rock was encountered at the shaft. As this material was practically a mud, the drift had to be heavily timbered and lagged, and the fault zone can not therefore be fully examined. It is certain, however, that it dips to the east, that it separates dacite on the east from the oldest andesite on the west, and that it represents a zone of considerable displacement accompanied by brecciation of the adjacent rocks. It is probably the Columbia Mountain fault.

The throw of the fault is not yet determined, but its maximum is certainly over 500 feet and may prove to be of twice that magnitude. It antedated the deposition of the lake beds and is thus older than at least a part of the doming of the region.

After the lacustrine epoch the region was domed, but the elaborate faulting which followed the deposition of the Siebert tuff at Tonopah had no representation in the Goldfield region. Some faulting occurred after the eruption of the olivine basalt, as may be seen south and west of Goldfield (Pl. II), but these faults are not of great structural importance. The fact that the remnants of the olivine-basalt flows now stand at different altitudes and occasionally slope at angles somewhat steeper than that of the broad mesa southwest of Goldfield suggests that there probably has been some differential movement or warping of the region since the basalt was erupted.

ORE DEPOSITS.

GENERAL CHARACTER.

The deposits hitherto of most importance in the Goldfield district are irregular bodies of rich oxidized gold ore. These have been derived from sulphide ores, of more or less complex mineralogical character, in which pyrite is accompanied by minerals containing copper, silver, antimony, arsenic, bismuth, tellurium, and other elements. The most notable features of the oxidized ore bodies are their remarkable richness and their equally remarkable irregularity. The ores are almost without exception associated with craggy outcrops of silicified volcanic rock, although only a very small proportion of these outcrops, which are extraordinarily numerous and constitute the most striking superficial feature of the district, have been found productive. Associated with the silicification other processes of locally intense alteration have also been active, producing in many cases a softening of the rock affected and thus serving to accentuate the silicified portions under the selective action of erosion. The deposits have formed along zones of fissuring which for the most part are very irregular in trend, are rarely traceable in any one direction for more than half a mile, and are not planes of notable faulting. Branching and intersection are very common. Many of the outcrops show no linear character, being mere irregular knobs of siliceous material. Some of these probably represent silicification at the point of intersection of two or more inconspicuous approximately vertical fissures. Others are erosion remnants of nearly horizontal silicified zones and, as in the case of Black Butte, merely cap the hills whose summits they form.

MINING DEVELOPMENT.

The principal producing mines in the Goldfield district in the latter part of 1905 were the Combination, Florence, January, Redtop, Sandstorm, and Kendall. The Jumbo, one of the most important properties in the district, was involved in litigation at this time and had produced no ore since the early part of the year. The Quartzite mine at Black Butte, now part of the Black Butte Consolidated property, was also practically idle, although it had shortly before attained a considerable production. The Reilly lease, although on the Florence ground, was worked as a separate mine, and added about $475,000 to the production of the district in less than two months' time.

The Combination Mines Company owns 10 claims and 3 fractional claims in the district, six of the claims forming an undeveloped outlying group about a mile east of the mine. At the time of visit the mine was worked through a steeply inclined shaft to a vertical depth of about 280 feet. There are 5 levels, the first a short adit 30 feet below the collar of the shaft and the others at intervals of 50 feet. The bottom level is under water. A new vertical shaft, equipped with a counterbalanced cage and a 30-horsepower electric hoist, was nearly ready for use at the end of the year. The levels run approximately northwest and southeast and the total length of ground explored is nearly 1,000 feet. On the northwest there is a connection with the January mine and on the southeast with the Reilly slope. The mine is equipped with a 20-stamp mill and cyanide plant. A tube mill was also being installed at the time of visit. Water for milling and other purposes is pumped from Alkali Spring, 10 miles northwest of the mine. The mill is driven by electricity, with steam as reserve power.

The Florence mine has a vertical shaft 350 feet deep and 6 very irregular levels. Most of the levels are at intervals of 50 feet, but between the 250-foot and 350-foot levels there is merely a short sublevel not connected with the main shaft. The general trend of the workings is about north-northwest, and the total length of exposed ground at the time of visit was, about 500 feet. Drifts were being rapidly extended northwestward, and possibly before this report appears the main workings will have been connected with the Reilly levels.

Hoisting at the Florence is done by electric power. The Reilly shaft, which lies between the main Florence shaft and the Combination shaft, is 210 feet deep, with four levels approximately 50 feet apart. These, like the Florence levels, are very irregular.

The Jumbo shaft was 450 feet deep at the time of visit and sinking was in progress. The shaft is west of the ore, and there were no crosscuts or drifts below the 250-foot level. Above this the rich ore of the lode, which is unusually regular for this district, has been stoped by lessees through half a dozen shafts for a length of about 600 feet. A mill of two Nissen stamps was in process of building in December, 1905.

The January mine has a vertical shaft 280 feet deep, exclusive of sump, with six levels at irregular vertical intervals. All the levels are sharply curved at the shaft and run from that point about 100 feet north-northeast and 160 feet southeast, both arms of the curve running toward the Combination ground.

The Redtop mine at the time of visit was in an early stage of development, the shaft being 180 feet deep, with levels at 100 and 160 feet below the surface.

The Sandstorm mine, being worked partly by the company and partly by lessees, is rather unsystematically developed. The greatest depth attained at the time of visit was 150 feet.

The Kendall shaft is about 200 feet deep, with levels at 60 and 100 feet below the surface.

Until very recently gasoline engines of from 12 to 30 horsepower have been almost universally used in the district for hoisting and other purposes, the use of steam at the Combination and January mines being the only exception to this practice. Electric power, however, is now coming into use. The electricity is generated on Bishops Creek, 98 miles west-southwest of Goldfield, at the east base of the Sierra Nevada and is supplied by the Nevada Power, Mining and Milling Company to both Tonopah and Goldfield. The present capacity of the main line is about 4,000 horsepower. The two districts together were using about 1,500 horsepower at the close of the year 1905. The rates in force at this time for continuous service, based on peak loads measured on the motor side of transformers, were as follows:

Rates for electric power per horsepower per month.

	Net rate.
1 to 10 horsepower	$20.00
11 to 20 horsepower	15.00
21 to 50 horsepower, $15 less 10 per cent	13.50
51 to 75 horsepower, $15 less 15 per cent	12.75
76 to 100 horsepower, $15 less 20 per cent	12.00
101 to 200 horsepower, $15 less 25 per cent	11.25
201 horsepower and upward, $15 less 30 per cent	10.50

Users of the power were required to furnish transformers and other connecting apparatus.

The present water supply of Goldfield is obtained from a well 90 feet deep, situated about three-fourths of a mile southwest of the center of town in the lake beds that underlie the volcanic capping of the mesa. The supply in September, 1905, was about 15,000 gallons a day, but experience with shallower wells at this locality has shown that the available water at any given depth tends to diminish, so that a continual supply can be secured only by occasionally deepening the well or by drifting through the water-bearing strata. The water supply of Columbia is obtained from Rabbit Spring. The general surface of the ground water of the district is not so deep as might have been expected in so arid a region. The 280-foot level of the Combination mine is under water, and the January mine has water at the same depth. This water, however, is entirely unfit for use in steam boilers. Near Diamondfield the Spokane shaft affords a good supply of water at less than 100 feet, and the Dewdrop or Highland shaft is in water below the 100-foot level. The water from the Dewdrop shaft is pumped to Columbia Mountain for use in milling. Most of the custom mills in the district derive their water from individual wells up to 200 feet in depth, sunk in the lake beds west of town.

An adequate quantity of water can probably be obtained for many years from wells and from the lower levels of the mines. Extensive and long-continued pumping operations, however, are likely to bring to the surface more water than is supplied by precipitation, so that a slow subsidence of the general level of underground water may be expected.

MILLING AND SHIPMENT.

According to an estimate made by Mr. Francis A. Thomson, superintendent of the New Western Reduction Company, and published in the Goldfield Sun of September 14, 1905, the average daily production of the Goldfield mines at that time was 70 tons of shipping ore of an average value of $200 a ton and 100 tons of milling ore at an average value of $40 a ton. In general, ore of lower grade than $100 a ton is milled in the district. The Combination Mines Company mills all of its ore except that exceeding $600 a ton in value, which is shipped.

The freight rates on ore to San Francisco, Salt Lake City, or Denver in 1905 ranged from $8 a ton for ore under $25 in value to $22 for $300 ore. For richer ores, from 3 to 5 per cent of the value per ton in excess of $300 was added to the $22 rate. For example, on $2,000 ore the freight would be $22 plus 5 per cent of $1,700, or $107 a ton. Smelting charges were $10 a ton on all siliceous ores, with an additional charge of $1 a ton if the silica were unusually high.

With the exception of the Combination, none of the mines had mills in operation at the end of 1905. The other mines either shipped their ore directly or disposed of it to one of the four reduction or sampling companies in the district. Some ore of lower grade has also accumulated on the various mine dumps.

The New Western Reduction Company pays 85 per cent of the assay value for ore, subtracting a charge of $12 a ton for milling. Improvements in this plant were in contemplation at the time of visit and these charges may since have been lowered. The Nevada Goldfields Reduction Company, popularly referred to as Frank's mill, buys ore outright and mills it or, in the case of ore from the January mine, pays 95 per cent of the assay less $12 a ton for treatment. The American Milling and Water Company buys ore or mills it at an average charge of $12 a ton. The Columbia Sampling and Ore Company acts as sampler or purchaser only, shipping the ore out of the district for treatment. The total maximum capacity of the three custom mills running in 1905 was about 70 tons a day. The milling facilities have probably since been increased by a new custom mill at Columbia, by the new Jumbo mill, and by improvements in the older plants. The Combination mill at the end of 1905 had just been increased to 20 stamps, with an estimated capacity of 70 tons.

DISTRIBUTION.

The rock alteration directly associated with ore deposition has affected all of the igneous rocks in the district that are older than the latite. The alaskite, however, shows less metamorphism than the older rhyolites, the two andesites, and the dacite, perhaps in part because of the comparatively small areas of this rock exposed. The most intense and conspicuous alteration is manifest in a belt which begins in the dacite south of Columbia and just east of the Columbia Mountain fault, and stretches in a slight curve eastward over Preble Mountain and for some miles east of the area mapped. In this belt the characteristic siliceous outcrops, largely in andesite, are more abundant and more conspicuous than in any other part of the district. There is a general tendency for the more persistent fissures to conform in strike to the trend of the zone, but the departures from this direction are so numerous and varied as to leave little more than a suggestion of system. A plan of the principal outcrops of silicified rock in the western part of this belt is shown in fig. 3. The exposures in this part of the district are better than elsewhere and probably over 90 per cent of the zones and centers of silicification are visible in outcrop. These

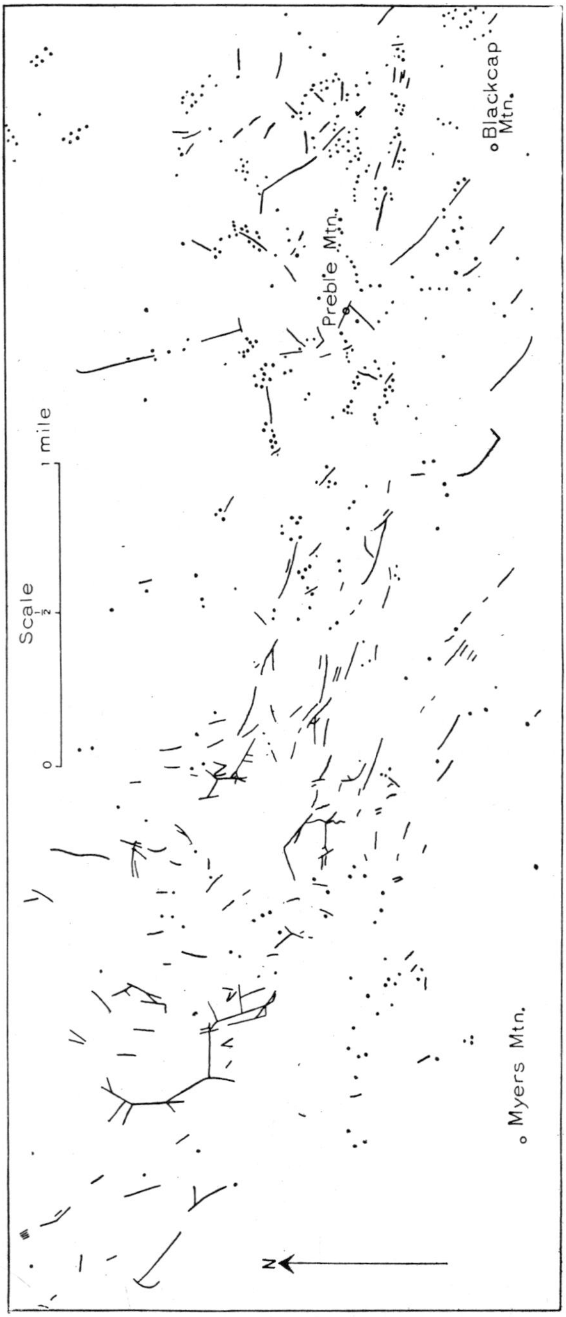

FIG. 3.—Plan showing distribution of siliceous ledges east of Goldfield.

outcrops have been plotted with a plane table. Their representation is necessarily conventionalized, as there is usually no distinct boundary between silicified and unsilicified rock and one center of silicification may merge into another. There is no attempt in the figure to represent relative size. A circular or irregular outcrop is represented by a dot and a linear or lodelike outcrop by a line. Many small outcrops are necessarily omitted, especially where the metamorphism has been most intense and the centers of silicification are closely crowded.

A second belt of alteration extends east and west across the northern part of the district from Kendall Mountain over McMahon Ridge and Black Butte. East of Black Butte the belt widens and becomes less distinct.

Between the two belts, siliceous outcrops are scattered rather irregularly over the district, in some places, as on Banner Mountain and near the Sandstorm mine, being clustered into local groups of some prominence or importance.

The distribution of ore is more restricted than that of the siliceous outcrops. The most important ore bodies thus far found are in dacite and are all within a small area, less than three-fourths mile square, lying just east of Goldfield and Columbia. The ore bodies of the Combination, Florence, Jumbo, January, and Redtop mines are all in this limited portion of the dacite mass, and those of the St. Ives, Simmerone, and Commonwealth mines occur farther east in the same rock. The rich ore shoots of the Sandstorm and Kendall mines are in the oldest effusive rhyolite. No important deposits have yet been found in the oldest andesite, although the occasional finding of small bunches of ore has encouraged prospecting in this rock, particularly in the area south of Columbia. The only large body of ore thus far opened in the younger andesite is that of the Quartzite mine at Black Butte. Most of the smaller properties north of Diamondfield are, however, in this andesite, which is also the country rock of the Gold Bar mine, 1 mile northeast of Myers Mountain.

No cause has been found for the preponderance of ore in the dacite. The two andesites have undergone similar alteration and do not differ greatly in petrographic character from the dacite. It is probable that the dacite is lithologically no more suitable for ore deposition than the andesites and older rhyolites, and future prospecting may considerably decrease the relative economic importance of this apparently more favored rock.

As the latite, lake beds, youngest rhyolite, and basalts show no evidence of siliceous alteration or of ore deposition, they are probably younger than the primary ores. This conclusion, however, is less definite than it might be had not these younger volcanic formations been eroded completely away from what appear to be the areas of most intense alteration.

FORM OF DEPOSITS.

There is no name in the literature of economic geology that is strictly applicable to the deposits of the Goldfield district. They can not properly be called veins, since in their formation fissure filling has had a wholly subordinate, in fact, almost negligible, part. In many cases they are not even approximately tabular in form and are not genetically related to any single main fissure. Walls are generally lacking except such slips as may have occurred since the original ore deposition.

In general, the deposits are irregular masses of dense flinty quartz formed by the silicification of fractured or shattered rock of various kinds by silica-bearing solutions. In describing them it will be necessary to distinguish between these masses as a whole and the bodies of ore that may or may not be associated with them. As the silicified masses form characteristic rough outcrops they may be conveniently referred to as *ledges*, ledge being a word in common use among miners for any outcropping vein, lode, or less regular deposit, while the ore bodies proper will be designated *pay shoots*. The pay shoots, almost without exception, occur in or near the ledges, but there are hundreds of ledges which, so far as known at present, do not contain a pound of ore.

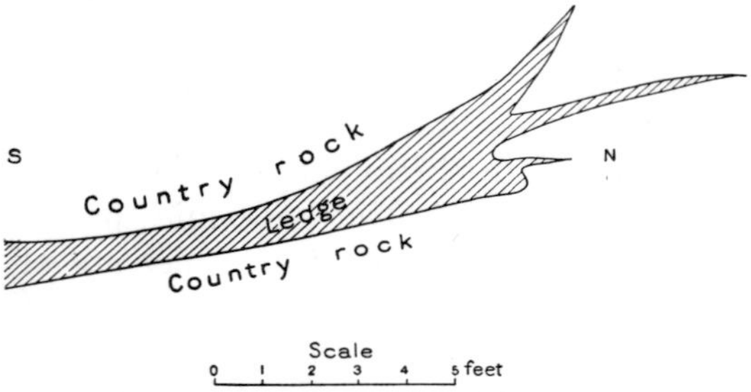

FIG. 4.—Vertical section through an irregular flat ledge as seen in the west wall of the Black Butte tunnel.

In plan, particularly as shown in outcrop, the ledges may be roughly circular, crescentic, rectilinear, or wholly irregular. Some of the circular or irregular masses are approximately horizontal remnants capping small knobs. Black Butte furnishes a good example of this type, a tunnel run through the hill having shown conclusively that the great silicified mass forming the summit of the butte is merely superficial and that similar but thinner zones of silicification traverse the kaolinized andesite beneath it. A section of one of these minor ledges as exposed in the wall of the tunnel is shown in fig. 4, the mode of branching and pinching being thoroughly characteristic of these deposits. Others may, as Mr. Spurr suggests,[a] have the form of pipes or chimneys, although mining operations have not yet demonstrated the descent of pipelike masses to any great depth. Many of the irregular ledges, such as miners would ordinarily term blow-outs, are situated at the intersections of lines of fissuring and silicification.

[a] Bull. U. S. Geol. Survey No. 260, 1905, p. 134.

The most important ledges, those upon which most work has been done and which have produced the greater part of the ore, are more or less linear or lodelike in plan. Notwithstanding the great irregu-

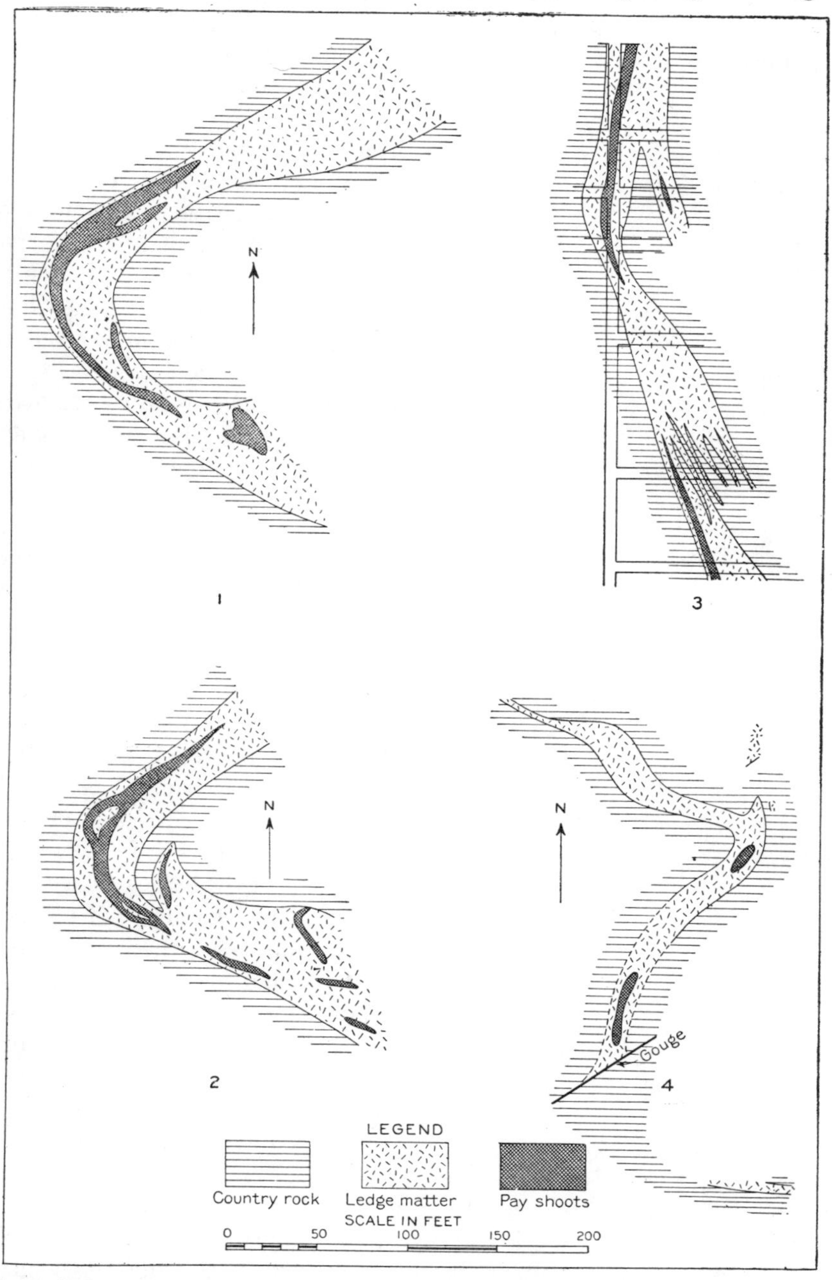

FIG. 5.—Ledges and pay shoots in the Goldfield district: 1, Plan of pay shoots and ledge on the first level of the January mine; 2, plan of pay shoots and ledge on the second level of the January mine; 3, diagrammatic cross section of the January ledge and pay shoots, looking north; 4, plan of pay shoots and ledge on the second level of the Florence mine.

larity of their associated pay shoots, the ledges of the Sandstorm, Kendall, Redtop, January, Combination, Jumbo, and Florence mines, as well as those of many less prominent properties, have this general lodelike character. The Sandstorm-Kendall ledge can be followed over the surface as a continuous outcrop of intensely silicified rhyolite for a distance of fully 1,500 feet. The Combination, Reilly, and Florence workings are all situated on a ledge of silicified dacite and andesite, which, although not quite continuous at the surface and somewhat curved, apparently represents one general zone of fissuring nearly half a mile in length. The Jumbo ledge has a fairly regular linear outcrop for a distance of 600 feet. The January ledge is sharply curved in plan, and the Redtop, owing to its width and the shortness of its outcrop, is less lodelike than the ledges of the other mines mentioned. Inspection of fig. 3 will show that a linear outcrop is not peculiar to the ledges of known economic importance, but is characteristic of many whose value is as yet unproved. The width of the ledges is variable, as may be seen from fig. 5, ranging in most cases from 10 to 100 feet.

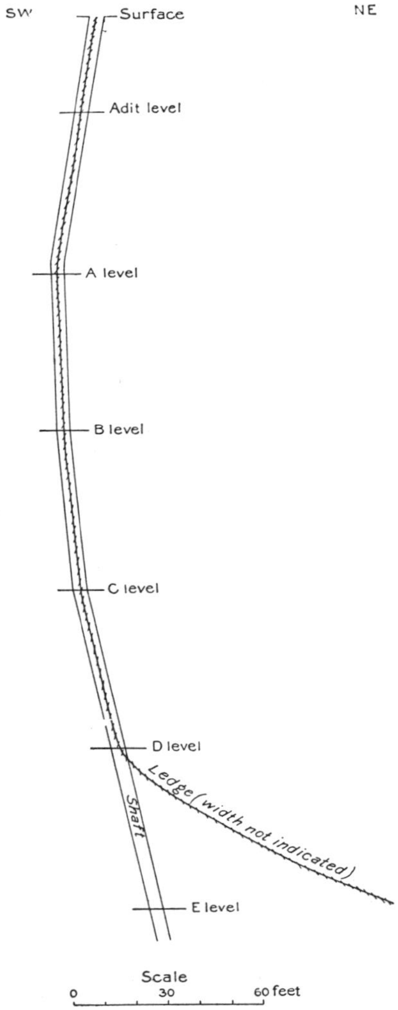

FIG. 6.—Cross section through the Combination mine, showing change in dip of the Combination ledge.

As none of the ledges has yet been explored below a depth of 400 feet, less is known of them in vertical section than in plan. Moreover, below the zone of oxidation the distinction between ledge and country rock in some cases becomes very indefinite. Oxidizing solutions, as a rule, penetrate the brittle, shattered ledge matter more readily than they do the soft unsilicified country rock. Consequently, within the zone of oxidation there is usually a remarkably sharp change in color and texture from the rusty siliceous material of the ledge to the gray, soft, pyritized country rock, locally referred to as porphyry. The lack of definiteness of some of the ledges at a depth of a few hundred feet is not entirely a matter of oxidation, however. In some cases silicification unquestionably decreases gradually with increase in depth.

Changes in dip are common. The Combination ledge has a dip of 81° to the southwest down to a depth of 80 feet. Below that it dips northeastward. At a depth of 230 feet the ledge, as shown in fig. 6, becomes very flat and exhibits little silicification. Unfortunately this part of the ledge, being under water, could not be seen at the time of visit. The Reilly ledge, which is continuous with what is known as the "No. 2 shoot" of the Combination mine, dips to the northeast for the first 160 feet below the surface, but then turns and dips in the other direction, as illustrated in fig. 7. The roll in the ledge is not regular or horizontal, as is shown by the fact that the levels of the Reilly workings, seen superposed in plan, are not parallel, but cross one another at considerable angles. This irregularity is the more remarkable, as the "No. 2 vein," in the Combination mine, contains one of the most regular pay shoots in the district. Similar changes of dip are very pronounced features of the ledge exploited in the main Florence workings. Although most of the levels are only 50 feet apart, a general plan of the mine shows them crossing and recrossing one another at angles of 30° or more, indicating abrupt changes in dip and strike.

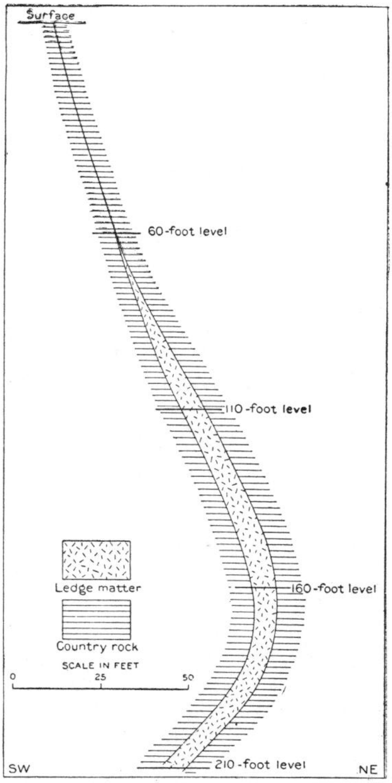

FIG. 7.—Cross section of the Reilly ledge, showing changes in dip.

In general, all the underground work done in the district shows that the ledges, although they have some lodelike features, are subject to such remarkable changes in dip and strike that methods of mining based on the behavior of ordinary veins and lodes are not applicable to them. Such eccentricities, as are shown in figs. 4 to 7, illustrate the uncertainties attendant upon underground exploitation and indicate that any attempt to carry development work far ahead of the known ore becomes largely a game of chance.

If the ledges are irregular, the pay shoots are even more so. In most cases these occur within and are completely inclosed by ledge matter; but there are some exceptions to this rule and some ore bodies are directly in contact with unsilicified country rock. Some of the pay shoots are mere bunches or "pockets." On the Velvet claim, for example, several thousand dollars' worth of rich ore was found at the surface, but efforts to find a continuation of this shoot in depth have thus far been futile. The St. Ives, Simmerone, and Conqueror (Tonopah Club) mines have had similar histories, although this in itself does not warrant the conclusion that deeper ore bodies may not yet be found in these properties. The Kendall mine at the time of visit had produced about $100,000, all from ore within 60 feet of the surface, and the deeper workings at that time showed no ore worth shipping. The rich ore of the Sandstorm mine has all come from very irregular pay shoots, which in many cases rest on flat floors of hard, barren, silicified rhyolite. These bodies are usually found at the intersection of later cross fissures with the main silicified ledge. Although some rich ore has recently been opened at a depth of 100 feet, the bulk of the production has come from ore within 50 feet of the surface. The ore bodies first worked in the Combination mine occurred in short irregular shoots, most of them at points of intersection of two or more fissure zones and with a tendency to a vertically elongated or pipelike form. A sketch plan of one of these shoots on level A is shown in fig. 8, and in fig. 9 are represented plans of shoots on the level below. At greater depth these irregular pay shoots appear to come together, and on level D the deposit is lodelike, with a length of about 250 feet. About 100 feet of this is shipping ore, the rest being of milling grade.

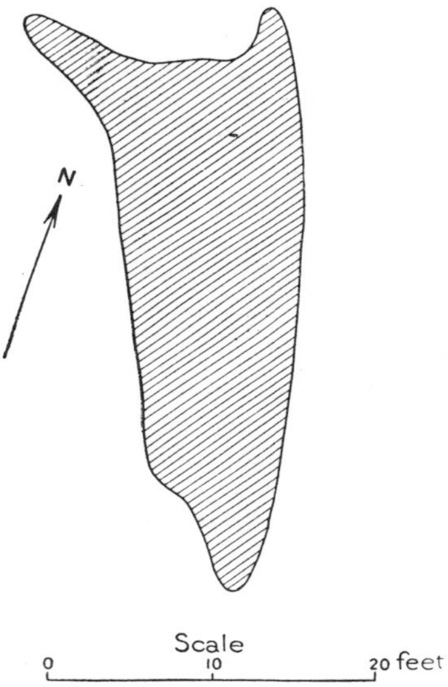

FIG. 8.—Sketch plan of pay shoot on A level of the Combination mine.

The so-called "No. 2 shoot" of the Combination mine is, with the possible exception of the Jumbo pay shoot, the longest and most regular body of ore known in the district, having been stoped on the 130-foot or B level (fig. 9) for over 300 feet. Moreover, it is separated by only a short, barren interval from the Reilly ore body, which is on the same line of fissuring. This pay shoot is the hanging-wall portion of a shattered silicified ledge, 30 to 40 feet wide, which strikes northwest and dips northeast at angles ranging from 75° to 80°. The ore, which is up to 6 feet in width, has no walls and can not always be distinguished by the eye from the barren or low-grade ledge matter. It lies near, but not always in contact with, a persistent gouge, which separates the oxidized silicified ledge matter from soft kaolinized and pyritized dacite. Although this pay shoot is so regular and persistent on the B level it apparently extends to the surface at only one point as the pipelike shoot of unusually rich ore which was followed in the No. 3 prospect shaft. Below this level the relations of the No. 2 pay shoot to the original Combination pay shoot are yet to be determined. As shown in fig. 9, the two shoots probably come together, although, as already pointed out, no constancy of direction can be safely assumed in this district, even for so apparently regular a ledge and pay shoot as the No. 2 of the Combination mine, especially as the same ledge in the Reilly workings is subject to remarkable changes in dip. In fact, attempts to find the No. 2 shoot on the D level of the Combination mine have thus far been unsuccessful.

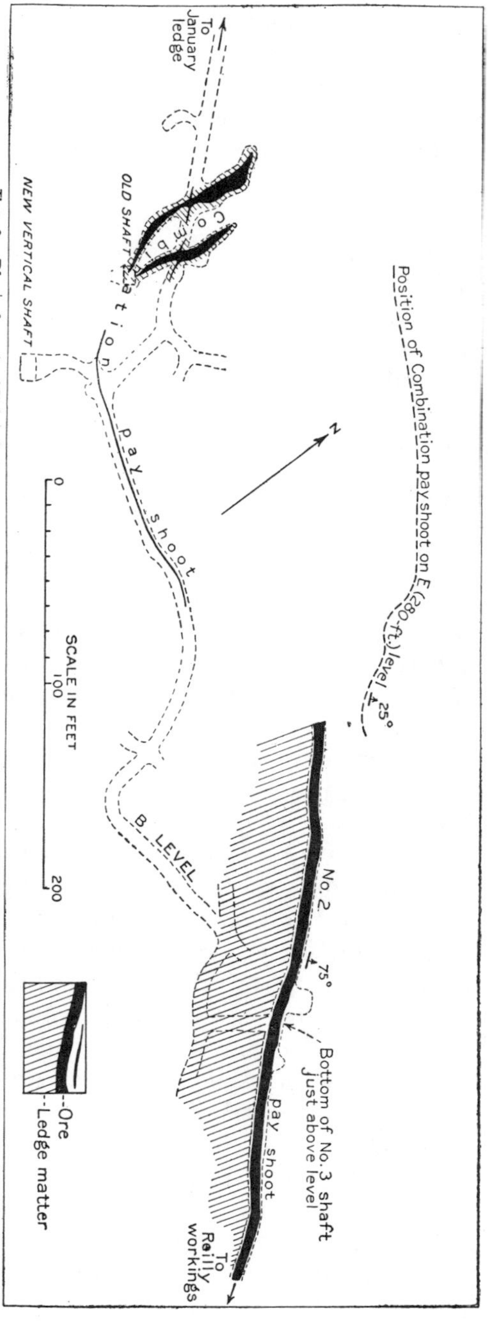

FIG. 9.—Plan of part of B level, Combination mine, showing diagrammatically the relations of the pay shoots.

In the January mine the erratic distribution of irregular pay shoots within the curved ledge is well shown in fig. 5. The two levels chosen for illustration are only 50 feet apart, yet the distribution and outlines of the ore bodies are notably different. The distribution of pay shoots in the Florence mine (fig. 5) is also without recognizable plan. The main ore body, from which the lessees obtained most of their ore, lies mainly between the 200-

foot and 150-foot levels, and the stope opened in it is the largest in the district. This pay shoot is a rudely lenticular mass dipping from 35° to 40° to the southwest. The stope is fully 75 feet long and up to 18 feet wide. The ore continues for a short distance below the 200-foot level, steepening to a dip of 70°. It consists of much shattered dacite (or possibly andesite), partly silicified, full of fine pyrite and traversed by countless seams of kaolin or similar soft white material. It grades into ledge matter without any definite walls.

MINERALOGY.

The unoxidized ore of the district, as exemplified in the Combination, Florence, Jumbo, and January mines, consists of pyrite, bismuthinite, and a reddish-gray copper mineral with the general composition of tetrahedrite. Native gold is usually associated with these minerals and in the rich ore may be easily visible. These minerals are at some places arranged in successive layers or crusts and in others are irregularly dispersed through a siliceous gangue, which usually shows evidence of being shattered and silicified rock.

The brown copper-bearing mineral contains antimony and sulphur and will be provisionally referred to as tetrahedrite. The absence of crystal form, however, makes it impossible to make a definite determination, as some rarer minerals, such as famatinite, have a chemical composition similar to that of tetrahedrite but differ in crystallization. Whether the tetrahedrite from Goldfield contains notable quantities of silver or gold has not yet been determined. It is usually an indication of rich ore.

Bismuthinite, the sulphide of bismuth, occurs as slender lead-gray prisms and needles and is regarded by the miners as a certain sign of good ore.

The richest unoxidized ore of the Combination, January, and Florence (including the Reilly) workings consists of free gold, usually in particles so fine that the auriferous portions of the ore look, at first glance, as if they were stained with ocher. The gold is intimately associated with tetrahedrite and bismuthinite. Pyrite is usually present also, but there is apparently no relation between the abundance of this mineral and the richness of the ore.

Concentric shells of ore minerals about silicified rock fragments are rather characteristic of the rich ore. Specimens from the Combination mine show an inner zone of free gold and quartz, up to an eighth of an inch in thickness. This is covered by a shell of tetrahedrite and this by an outer crust of radial iron sulphide, probably pyrite and not marcasite. In the Florence mine similar fragments show, first, a shell of pyrite, then one of tetrahedrite, and, finally, a thick crust of quartz speckled with native gold, tetrahedrite, and pyrite, and transfixed by needles of bismuthinite. Some of the particles of gold are embedded in the compact quartz, others are inclosed in the bismuthinite. The different crusts are not in every case sharply defined nor are they necessarily continuous.

Tellurides occur sparingly in the Goldfield ores. A so-called telluride ore from the Reilly lease was found by Dr. W. T. Schaller, of the chemical laboratory of the Survey, to contain only a trace of tellurium. A dark-gray mineral occurring in small specks in compact white quartz and silicified andesite, in the Jumbo Extension mine near Diamondfield, is a telluride of gold, but the mineral is not sufficiently well crystallized for satisfactory determination of its species. As Dr. Schaller reports the absence of silver, the telluride can hardly be petzite. A similar mineral from the Goldfield-Belmont mine, three-fourths mile north of Diamondfield, contains gold and tellurium, but no silver or lead. The material, however, is not sufficiently abundant or well enough crystallized to permit specific identification. Tellurium has also been detected by Dr. W. F. Hillebrand in ore collected by Mr. Spurr[a] from the Combination and January mines.

A small speck of sphalerite was found by Mr. W. H. Blackburn in the Goldfield-Belmont mine, but galena has not been noted in any of the mines so far as known. Chalcopyrite, a common mineral in most mining districts, was not found at Goldfield in the course of the present investigation, but it is probably not entirely absent, and is said to occur sparingly in the Sandstorm and Florence mines.

The common gangue of the unoxidized ore is quartz. This is usually compact, almost flinty in texture, and in most cases bears unmistakable evidence of having resulted from the silicification of dacite, rhyolite, or andesite. Large vugs and conspicuously crystalline quartz, such as are found in typical veins all over the world, are practically absent from the Goldfield district, where the free development of quartz crystals in open spaces is represented only by drusy films lining pores left by the solution of phenocrysts or incrusting small interstitial cavities in brecciated material.

Associated with the quartz in much of the ore are soft white substances, such as are usually termed "talc" by the miners. In some cases the white material is kaolinite (hydrous silicate of aluminum), in others it is sericite (silicate of aluminum and potassium), but one interesting result of the examination of the so-called "talc" is that a large part of it is composed of the mineral alunite, a hydrous sulphate of aluminum and potassium. Gypsum, while not known with actual sulphide ore, does occur crystallized with quartz and pyrite below the zone of oxidation in the Goldfield-Belmont mine.

No sulphantimonites or sulpharsenites of silver were noted in the mines at the time of visit. A specimen of rich ore from the Florence mine, in the possession of Mr. F. O. Altinger, an assayer in Goldfield,

[a] Ores of Goldfield, Nev.: Bull. U. S. Geol. Survey No. 260, 1905, pp. 137 and 138.

showed, however, a small speck of ruby silver (proustite) and some tabular crystals which were apparently polybasite or stephanite.

Cinnabar is distributed rather abundantly through a mass of intensely altered andesite about 2 miles east of Black Butte in crystalline particles and aggregates up to an inch in diameter. The deposit has not been developed, but appears to be at least 150 feet long and 50 feet wide, so that it may prove to be of commercial importance. Native sulphur occurs in a similar manner, but in bunches up to 6 inches in diameter, in a bleached andesitic breccia about a mile east of Tognoni Springs. Small quantities of sulphur have also been found in the croppings of the Blue Bull mine. It is doubtful whether the mineral is abundant enough to be of economic importance.

Oxidized ores have supplied a large part of the gold produced, and in some mines, particularly the Sandstorm and Kendall, no sulphide ores are as yet known. As a rule the oxidized ore is a soft, shattered, more or less earthy material, usually stained brown by oxide of iron. Fragments of rusty porous quartz are mingled with kaolin, alunite, gypsum, alum, oxide of iron, and various earthy mixtures of no definite mineral composition. Some of the rich ore, as in the Sandstorm mine, is a nearly white, impure kaolin, gritty from the presence of minute crystals or grains of quartz and containing all through it abundant specks of free gold. Barite, in crystals up to an inch in length, is abundant in this ore. In the Redtop mine some of the richest ore is a firm gritty ocher, of bright yellow color, containing abundant very fine particles of gold. In January and Combination mines some of the most characteristic rich ore consists of porous rusty quartz (silicified dacite), in which the pores and crevices are partly filled with yellow, earthy limonite and tiny pearly scales of bismite, the oxide of bismuth. In some specimens these scales form pseudomorphs after bismuthinite. They are usually a sign of rich ore, but are occasionally found in low-grade or barren material. Native alum, or kalinite, and alunite are rather abundant in the oxidized ore of the January mine. In the mines at Black Butte the richest oxidized ore usually shows little greenish-yellow specks of a ferric tellurite, which Dr. W. F. Hillebrand has determined to be either emmonsite or durdenite. The presence of this mineral shows that the original sulphide ore contained one or more tellurides. Tellurite, a honey-yellow crystalline mineral, which is the oxide of tellurium and should not be confused with the salts of tellurous acid known to chemists as tellurites, was noted in a partly oxidized telluride ore from the Goldfield-Belmont mine near Black Butte. Melanterite (hydrous ferrous sulphate), in some cases mixed with a very little chalcanthite (hydrous copper sulphate), occurs in the partly oxidized ore of the Combination and Florence mines.

The valuable constituent of the oxidized ores is native gold. Native silver or halogen compounds of silver have not, so far as known, been recognized in the Goldfield district. The gold, which has undoubtedly undergone some concentration in the process of oxidation, is in most of the ores very finely divided, and even in the richest ores might easily be overlooked by one not accustomed to its appearance. Some of the Sandstorm ore on close examination resembles a fine, somewhat rusty, sandstone having gold as the cementing material.

GENESIS OF THE ORES.

Much office and laboratory work remains to be done before the genesis of the Goldfield deposits can be satisfactorily discussed. A few suggestions bearing upon this problem are all that can be offered at this time.

The formation of the ledges was initiated by extensive fissuring of all the rocks of the region with the probable exception of the latite and younger formations. The fissures formed a complex network, in which there is no obvious system, and although the disturbance was of a kind to produce broad zones of cracked and even shattered rock it did not in general dislocate the formations by persistent or structurally important faults. This network of irregular fissures furnished channels more or less devious to hot ascending waters carrying silica, hydrogen sulphide, or alkali sulphides, and possibly sulphurous or sulphuric acid. The waters held in solution also gold, silver, bismuth, copper, antimony, arsenic, and other metals, as well as small quantities of tellurium. A large part of the pyrite associated with the ores was undoubtedly formed by the action of sulphur-bearing solutions on the ferruginous silicates of the country rock, but some of the iron was probably brought up in solution also. The effect of these waters was to convert large masses of the fissured rocks to quartz—rhyolite, andesite, and dacite all being altered by molecular replacement to siliceous masses very similar in general appearance and in places where the alteration was most intense retaining but faint traces of original igneous texture. Other portions of the same rocks were changed to soft aggregates of kaolin, sericite, alunite, and pyrite, the two kinds of alteration apparently going on side by side. The practical absence of carbonates from the ore deposits is noteworthy.

It has been a common practice in the district to refer to the silicified portions as dikes of rhyolite and to the soft, pyritized country rock as porphyry. This is misleading, as ledge and country rock were in all cases originally the same eruptive rock.

The ledges are complex structures. The first silicification seems in no instance to have produced the ore as mined to-day, but was followed by at least one other period of fissuring and brecciation, as is shown by the fact that the rich sulphide ore often coats fragments that are themselves veined and silicified.

How far this rich ore was deposited by ascending solutions and how far by solutions that percolated downward through a zone of oxidation is a question still under investigation. It is probable that to a large extent the ores were the result of deposition from rising solutions.

Another question of great practical importance relates to the changes in the ore with increase of depth. There is considerable evidence in favor of the conclusion that ores of the grade now required for shipment are not likely to extend to as great a depth as 1,000 feet. Available workings certainly show a tendency of the sulphide ores to decrease in value below a depth of 300 feet. It should be said, however, that the mines as a whole are too shallow to establish this tentative conclusion as a fact.

VALUE OF THE ORES.

As a whole the ores thus far mined at Goldfield are unusually rich. Prior to January 1, 1905, the Combination mine shipped a total of 1,166 tons of ore having an average assay value of $419 per ton. This ore averaged 20.22 ounces of gold and 2.68 ounces of silver per ton. The best ore from the mine was taken from the small No. 3 shaft, which afforded about 309 tons valued at $159,000, or $514 per ton. Approximately 59 tons of this averaged, according to Supt. Edgar A. Collins, 54.8 ounces of gold per ton, which is equivalent to a value of nearly $1,100. This ore, which was oxidized, contained only 1 ounce of silver to each 20 ounces of gold.

Mr. J. E. Spurr[a] estimated in 1904 that the average value of the ores mined in the Goldfield district was from $200 to $300 per ton or more. In 1905 the average value of shipping ore was about $200 per ton and of milling ore from $40 to $50 per ton.

Shipments averaging over $1,000 per ton are by no means uncommon. One of 14¼ tons from the Kendall mine is said by Mr. Spurr[a] to have yielded $45,783 net in the preliminary milling, while the tailings still contained about $1,000 a ton. A shipment of sulphide ore from the Florence mine, made in 1905, just before the date of visit, averaged 100.8 ounces of gold, 3.5 ounces of silver, and 3.5 per cent of copper, giving a gross value of over $2,000 per ton. The silver in this shipment was unusually high for an ore from this part of the district. An analysis of a sample of Florence ore published in the Goldfield News of April 7, 1905, is as follows:

Analysis of sample of Florence ore.

Gold	ounces per ton	70.21
Silver	do	7.8
Copper	per cent	6.9
Iron	do	28.5
Zinc	do	.3
Arsenic	do	.3
Sulphur	do	13.20
Silica	do	43.44

[a] Loc. cit., p. 136.

This analysis is evidently neither accurate nor complete. If, as is probable, the copper was present as tetrahedrite, there should be considerable arsenic or antimony in the ore. The figures cited, however, in connection with those previously given, serve to indicate that the silver is largely combined with the copper in the mineral tetrahedrite.

Some extremely rich ore was obtained from the Reilly lease, occasional specimens several inches in diameter being nearly one-half gold. Several shipments from this lease are said to have assayed $20,000 a ton, and, according to Mr. Frank Oliver, one lot of over 4 tons was worth from $6,000 to $7,000 a ton.

Some of the ore from the vicinity of Diamondfield carries a greater proportion of silver than that from the mines near Goldfield. A specimen containing tetrahedrite, taken from the Quartzite mine of Black Butte, yielded on assay 11.06 ounces of gold and 115.70 ounces of silver per ton. Samples of ore with tetrahedrite from the Goldfield-Belmont mine assay, according to Mr. W. H. Blackburn, yielded from 1 to 2 ounces of gold and from 30 to 60 ounces of silver per ton. A similar high proportion of silver is found in the ore of the Jumbo Extension mine. Some bunches of ore in the Blue Bull mine also show a very high proportion of silver, one assay giving three hundred and eighty-two times as much silver as gold.[a]

Although, as is to be expected in a new district, much of the very rich ore has been oxidized, yet some of the sulphides in the mines near Goldfield have exceeded in value anything found in the oxidized zone above them.

MINE MAPS.

Very few of the mines at Goldfield have satisfactory maps of their underground workings, and it is consequently in many cases difficult to understand the relation of one ore body to another in the same ledge or even to ascertain the shape and dimensions of individual pay shoots. This condition will doubtless be improved as the mines become older and as work is systematized, but the present development work in the larger mines could undoubtedly be laid out to better advantage if good plans of the levels were supplemented by numerous accurate sections and if important fissures, ledge matter, and ore were plotted upon the mine maps. Models constructed of sheets of glass, such as are used by the largest mines at Cripple Creek, would be especially useful as aids to understanding the relations of the ore bodies to one another in space, and if kept up to date and intelligently studied would doubtless furnish suggestions for exploration. Where ore bodies are so irregular in shape and distribution good maps and practical models are necessities rather than luxuries and cost less than many a crosscut driven at random into barren ground.

[a] Spurr, J. E., loc. cit., p. 137 (footnote).

BULLFROG DISTRICT.

INTRODUCTION.

In the autumn of 1905 a topographic map of the most important part of the Bullfrog district, covering an area about 7 miles from east to west and 2½ miles from north to south, was made by Mr. William Stranahan, topographer, on the scale of 1:24,000, with contour intervals of 20 feet. This map shows the town of Beatty near its eastern margin and just includes the Original Bullfrog mine on the west. Geological work was begun in December of the same year and completed in March, 1906. Messrs. W. H. Emmons and G. H. Garrey mapped the geological formations and studied in detail the relations and character of the rocks, while the writer devoted a shorter time to a general survey of the field and to the examination of the ore deposits.

The present account of the district is merely a preliminary sketch, in advance of a complete report, now in preparation.

SITUATION OF THE DISTRICT.

The name "Bullfrog district" is usually rather vaguely applied to a large tract of desert hills and mountains extending far to the north and west of the area particularly investigated. For convenience of description, however, the name as used in this report will refer in most cases to the area mapped, which includes nearly all the ore deposits thus far developed, although there are some scattered prospects of considerable promise lying without its boundaries.

The position of the district is shown in the index map (fig. 1) on page 9. It lies 60 miles south-southeast of Goldfield, with which it is connected by stage and automobile lines from 70 to 80 miles in length and about 20 miles northeast of the great trough of Death Valley. Railroads are in course of construction into the district from the south, one from Las Vegas on the San Pedro, Los Angeles and Salt Lake Railroad, and one from Ludlow, Cal., on the Atchison, Topeka and Santa Fe Railway, by way of Death Valley. Railway connection is projected, also, between the Bullfrog and Goldfield districts.

The principal town is Rhyolite, situated at the south base of a short east-west range of hills that connects the Funeral or Grapevine Mountains on the west with Bare Mountain and other irregular groups of peaks, ridges, and mesas on the east, and separates the Amargosa Desert on the south from a similar desert basin extending northward past Stonewall Mountain toward Goldfield.

The other towns in the district are Bullfrog, which adjoins Rhyolite on the south, and Beatty, 4 miles east of Rhyolite, on the so-called Amargosa River, a feeble trickle of water that is fed by springs a few miles north of the town and is finally absorbed in the sand and gravel of the Amargosa Desert.

HISTORY.

Long prior to the discoveries of ore at Tonopah and Goldfield the springs in Oasis Valley, just north of the site of Beatty, proved attractive to a few desert wanderers who took Indian wives and established rude ranches along the Amargosa River, the only stream of drinkable water within a radius of 40 miles. Prospectors traveling northeastward from Death Valley or northwestward from the springs at Ash Meadow would pass the night at one of these ranches, perhaps spend a few days examining the hills in the neighborhood, and then again brave heat and thirst on the shadeless plains.

The remarkable development of Tonopah and the finding of rich ore at Goldfield stimulated prospecting throughout all of southwestern Nevada, and the rapid advance of the material adjuncts to civilization into this most inhospitable region greatly lessened the hardship and danger that had hitherto attended a traveler far from bases of supply and vitally dependent upon springs, often long distances apart and not always readily found. In the autumn of 1904 the Bullfrog claim was located on a prominent outcrop of quartz, about 3 miles west of the site of Rhyolite. The name, which afterwards became that of the district, is said to have been suggested by the green color of the ore. The ironical humor of associating a paludinous fauna with scorching aridity is characteristic of the Nevada pioneers.

The usual rush of prospectors followed the first discovery, and most of the ground between the Bullfrog claim and the Amargosa River was soon covered with claims. A settlement of tents, known as Amargosa City, sprang up near the Bullfrog claim, but as the center of interest shifted eastward to Bonanza and Ladd mountains the first town was abandoned for the settlements of Bonanza, Bullfrog, and Beatty. In February, 1905, a town-site company laid out the streets of Rhyolite and, by offering a certain number of lots free, succeeded in a few days in establishing this as the chief town of the district. Whether it will remain so in spite of the inducements offered by other town-site companies and the proposal to carry the main railway lines past Beatty remains to be seen. It has, however, an advantage of position with regard to the mines that is much in its favor.

PRODUCTION.

Work in the Bullfrog district up to the end of the year 1905 had been confined mainly to prospecting and to blocking out ore. The Montgomery-Shoshone mine is said to have shipped about 100 tons of rich ore, the original Bullfrog about 13 tons, and the Denver at the beginning of the year 1906 had shipped 1,000 sacks. As all ore at the time of visit had to be hauled by wagons to Goldfield or Las Vegas, the general tendency among the mine owners was to delay shipping until the entrance of a railroad should afford better facilities.

TOPOGRAPHY.

The Bullfrog Hills attain a maximum elevation of 6,095 feet above sea, the highest point being about a mile beyond the northern boundary of the area covered by the detailed map. The hills, which are steep, rocky, and practically bare of vegetation rise sharply from the gently sloping desert plains that border them on the north and south. The marginal elevation of these plains ranges from about 3,300 feet on the Amargosa River at Beatty to about 4,500 feet at Mud Spring 4 miles north of Rhyolite, and the local relief of the hills is thus about 2,800 feet. Southeast of Beatty, Bare Mountain, which culminates in a peak 6,235 feet high, presents a steep southwest front to the Amargosa Desert and an eastern face almost as precipitous to an embayment of the same desert. On the north, however, the mountain, which is composed principally of schist, limestone, and quartzite, merges into hills which, in spite of the topographic break due to Oasis Valley, are structurally and lithologically an eastern continuation of the Bullfrog Hills. To the west, the Bullfrog Hills extend without any definite topographic or geological break into the Grapevine Mountains.

The hills in the area particularly described in this report constitute a very irregular group, partly buried by the thick alluvium characteristic of the broad, undrained basins of arid Nevada. The upper limit of the alluvium, unlike that at Goldfield, is remarkably definite, and above it the rocky structure of the hills is laid bare to a degree highly satisfactory to the geologist. A view over the district from any high point suggests at once that the panorama spread out before the observer owes its main features to the cooperation of faulting and erosion on a thick series of lavas. Many of the ridges are obviously fault blocks, most of them tilted to the east, so that their western scarps show the horizontal banding of successive flows while their northeast faces exhibit merely the irregular pinnacles and ravines characteristic of the erosion of a dip slope of a single rock layer of generally homogeneous texture. Moreover, as one follows with his eye along one of the ridges the individual white, gray, green, pink, brown, or black bands that represent the edges of different flows, he finds that they can never be traced far before they are sharply offset or cut off entirely in a manner that can be due only to faulting. Thus it is evident that the faults are not parallel but intersect one another, and the irregularity of the relief, the absence of continuous lines of parallel ridges and valleys, is seen to be the visible expression of a complexly faulted structure. How far this expression is direct and how far it has been modified or transformed by erosion will be considered in another place. There are no perennial streams in the district except the Amargosa River, but erosion, though fitful in its activity, proceeds at times with great energy, and the water that falls upon the hills undoubtedly accomplishes more erosive work than would the same quantity under the conditions prevailing in a humid country well covered with vegetation. The arid, stony slopes of the Bullfrog Hills are elaborately carved by running water, and the so-called "box canyon" a mile west of Rhyolite is striking testimony to the power of a stream that is probably active for only a few hours or a few days in the year.

GENERAL GEOLOGY.

INTRODUCTION.

Although the writer has personally supervised the general geological work in the district and has himself thoroughly traversed the area, most of the facts in the following brief description of the general geology are derived from the notes of Messrs. Emmons and Garrey, to whom credit should be given for the actual working out of the intricate faulting and for the discrimination and correlation of the various lithological units. Their complete geological map and a more thorough discussion of the geology than is here possible will form parts of the final report on the district, which is now in preparation. They are not directly responsible for this preliminary outline, some of the statements in which may require modification as the material gathered in the field is more thoroughly studied.

THE ROCKS.

The prevailing rocks of the area studied—the ones that give character to the topography and that contain the ore deposits—are rhyolitic flows, presumably of Tertiary age. Much older rocks, however, are exposed in the southwestern and southeastern parts of the district and these will first be briefly described.

Three miles west-southwest of Rhyolite, south of the stage road (Pl. III), is a little group of low, rounded, whitish hills, composed of schistose and gneissic rocks cut by numerous irregular dikes of pegmatite. Some of the schists are practically quartz schist. Others are quartz-muscovite schists, and there are a few bands of amphibole schist. At one place there is a small exposure of dolomitic limestone, which appears to be an integral part of the crystalline formation, although it is not notably metamorphosed. It is possibly a small residual of a limestone bed that once covered the schistose rocks. Intimately associated with the schists are bands of biotitic gneiss with drawn-out, eyelike feldspars. The pegmatite dikes are very numerous and very irregular. In most of them quartz is more abundant than feldspar, and no essential difference could be detected between the white vitreous quartz that forms part of pegmatite dikes and the similar quartz that, without the accompaniment of feldspar, forms bunches and irregular lenticular veins in the gneiss and schist. Both carry a little pyrite. The metamorphic rocks are cut by a dike of rhyolite, probably once connected with one of the flows to the north, and by two varieties of dark hornblendic porphyry.

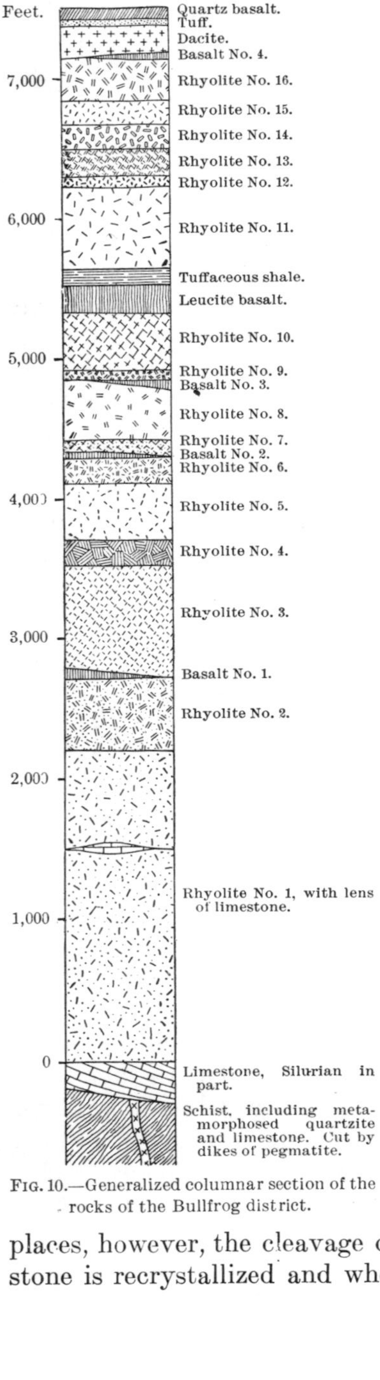

FIG. 10.—Generalized columnar section of the rocks of the Bullfrog district.

North of the stage road and a short distance west of the Original Bullfrog mine, the schists are overlain by gray limestone. Fossils collected by Mr. Garrey from a similar limestone about a mile west of the area mapped are regarded by Mr. E. O. Ulrich as of Silurian age. The schists are thus probably pre-Silurian.

For the distance of nearly a mile east of the Original Bullfrog mine the rocks are concealed by alluvium, but on the eastern edge of this alluvial embayment a little of the schist and limestone is exposed. East of this exposure the alluvium of the Amargosa Desert laps up over the Tertiary volcanic rocks.

About 4 miles east of Rhyolite and due south of Beatty, schistose rocks are again exposed between the alluvium and the rhyolite, forming relatively low hills in the southeast orner of the district and extending eastward into Bare Mountain, which affords magnificent sections of their structure. These rocks are clearly metamorphosed limestones, shales, and quartzites. The schistosity is generally parallel with the beds, which are bent into gentle folds and somewhat faulted. In a few places, however, the cleavage crosses the bedding planes. The limestone is recrystallized and where originally it graded into calcareous shales it now shows a gradual passage from crystalline limestone through calcite schist into mica schist. The quartzite has in part been changed to quartz schist. The schists are cut by a few pegmatite dikes, but these are not nearly so abundant as in the southwest corner of the district. At one place bunchy lenticular quartz veins in the schist are cut by a pegmatite dike. Veins of white quartz are fairly abundant, but are short and irregular. Some of them carry a little galena, but no other sulphide was noted.

In Bare Mountain the schists are apparently overlain by thick light-gray limestones, which are not particularly metamorphosed and are probably younger than the thin-bedded yellow limestones in the schists. Whether the schists are of the same age as those west of Rhyolite is an open question, although such general evidence as is available seems to favor an affirmative answer.

The total thickness of the rhyolite flows, including a few relatively thin layers of basalt, is estimated at from 7,000 to 8,000 feet. The provisional columnar section (fig. 10) will give some idea of their sequence and relative thicknesses. As shown in this section, 16 different rhyolitic formations are distinguished and will be represented on the final geological map. These formations are not all individual flows. Some probably consist of several superposed sheets of rhyolite of similar lithological character. Others may be portions of flows which by reason of pronounced difference in color or texture are conspicuous elements in the landscape and are important aids in the investigation of the faulting which is the principal feature of the local structure. It is obvious that for the purposes of description and reference each of the rhyolitic units must receive some distinctive designation. The usual practice is to apply to each formation the name of the place where it is most typically developed or exposed, as, for example, the "Beatty rhyolite;" but even were there sufficient geographic names available, their application to so large a number of geological units of the same general kind of rock would heavily burden the reader's memory and would probably lead to confusion. It seems best, therefore, to number the various rhyolitic formations from the base upward. Four flows of feldspathic olivine basalt will be similarly distinguished from one another. Should detailed work in adjacent areas require the interpellation of additional flows between those here recognized, they might if it is desired to retain the numbering now proposed, be designated "rhyolite No. 2 *a*," "rhyolite No. 13 *a*," and so on. It will probably be long, however, before any such expedient will be necessary. The estimates of thickness indicated in the columnar section are rough approximations and are likely to be somewhat modified in the final report.

The basal member of the Tertiary volcanic series, rhyolite No. 1, is best exposed in and near the deep ravine locally known as the "box canyon," about 1¼ miles west of Rhyolite. It is here much cracked and jointed and weathers in rather rounded yellowish masses. Rhyolite No. 1 undoubtedly includes more than one flow and shows considerable variation in texture. Much of it is fine grained and slightly porphyritic, often showing abundant small spherulites in the groundmass. Such spherulitic varieties rarely contain phenocrysts of quartz. In places the rock grades into aphanitic glassy rhyolites in which the laminæ due to flowage are so thin and regular that slightly weathered exposures have a remarkably close resemblance to outcrops of shales. Other facies are regularly banded flow breccias which in places simulate coarse sandstones or grits. At one point, about 2½ miles northwest of Rhyolite, a lens of dark-gray limestone occurs in rhyolite No. 1, showing that this formation is not a single flow, but represents at least two eruptions separated by an interval within which limestone to a thickness of several feet was deposited in a probably local basin.

In addition to the large area in the vicinity of the "box canyon," rhyolite No. 1 is exposed on the western edge of the district near the Original Bullfrog mine and on the east side of the Amargosa River at Beatty. The maximum thickness of this formation is provisionally estimated at 2,200 feet.

Rhyolite No. 2, which directly overlies rhyolite No. 1, is a flow breccia, usually light gray, but in some places of dull purple or pinkish color, and consists of a glassy groundmass, generally showing lines of flowage, in which are embedded fragments of rhyolite and basalt. The fragments are generally less than 2 inches in diameter and in most cases are darker than the matrix. This rhyolite is well exposed on the hill just north of the Original Bullfrog mine and on the slope above Buck Spring. Its thickness is estimated at 500 feet.

Between rhyolites No. 2 and No. 3 a flow of basalt intervenes at a few places, particularly near Buck Spring, 2 miles northwest of Rhyolite, where the flow attains a maximum thickness of 40 feet. The rock is an ordinary olivine-bearing feldspar basalt and may conveniently be referred to as feldspar basalt No. 1.

Rhyolite No. 3, which forms the greater part of the hill between Buck and Sullivan springs, 2 miles northwest of Rhyolite, resembles in its lower part the flow breccia of rhyolite No. 2, but grades upward into a somewhat vesicular glassy rock with pronounced flow lines. Its thickness varies from 300 to 500 feet. None of this rock is exposed in the eastern half of the district, although it probably occurs below the present surface.

Rhyolite No. 4 is variable in color, the common tints being greenish-white, pale bluish-gray, pink, dull purple, light buff, and cream. It is of porous texture and rarely shows conspicuous crystals. This flow is about 200 feet thick in Bullfrog Mountain (U. S. land monument 170), north of the Original Bullfrog mine. It is exposed also on the north edge of the district northeast of Buck Spring, but is not known east of this point.

Rhyolite No. 5 is a characteristic and important member of the volcanic sequence. It forms the summit of Bullfrog Mountain, where it is approximately 500 feet thick. The west slope of Bonanza Mountain is made up of this rock, as is also the lower western slope of Ladd Mountain, just east of Rhyolite. The same rhyolite is the principal rock of the southern and eastern slopes of Sutherland Mountain, of which Bonanza Mountain is a southeastern spur. It is thus the country rock of the Denver mine.

The principal lithological characteristics of this formation are the presence of phenocrysts of black mica and the frequent occurrence of spherulites, some of which, on Sawtooth Mountain, are over 18 inches in diameter and others east of Beatty attain a diameter of 5 feet. The lower part of the formation is in most places a well-defined band of brown spherulitic or perlitic glass up to 50 feet in thickness, and a similar band occurs at some localities about 200 feet above the base of the formation, which has a total thickness of from 500 to 800 feet. In the vicinity of the Denver mine the rock is compact, shows very few porphyritic crystals, and is prevailingly buff in tint. Farther west, however, this variety grades into rather more typical facies.

Rhyolite No. 6 is in most places a nearly white, porous, somewhat pumiceous flow breccia, with occasional pale-greenish or pinkish tints. It is this formation that makes the prominent white band visible from Rhyolite on the slopes of Busch Peak and of Ladd and Sutherland mountains. It is also exposed along the western bases of many of the hills between Ladd Mountain and the Amargosa River. The thickness of rhyolite No. 6 varies from a few feet in Bullfrog Mountain to about 300 feet on the southwest slope of Montgomery Mountain.

One notable feature of rhyolite No. 6 is the occurrence within it of masses of basalt. Some of these are irregular. Others are thin tabular bodies which stand nearly vertical and which, did they occur alone, might easily be mistaken for dikes. Close examination of the contacts, however, shows that the rhyolite was fluid after the basalt had solidified. Little tongues of rhyolite penetrate the darker rock. The flow banding of the rhyolite, moreover, conforms locally to the surfaces of the basalt masses. That the apparent dikes are really inclusions is certain, but no satisfactory explanation has yet been found for their vertical attitude in a flow that must have had a generally horizontal movement or for the source of the basaltic material. These inclusions are best seen near the trail from Rhyolite to Sullivan Spring, where it passes through the saddle in the ridge south of Busch Peak.

East of Rhyolite the white No. 6 rhyolite is directly overlain by a very persistent and characteristic pink rhyolite, which is No. 7 of the rhyolitic sequence. In the ridge west of the town, however, a basaltic flow, feldspar basalt No. 2, intervenes between the two rhyolites. This basalt is well exposed on the trail from Rhyolite to the Eclipse and Denver mines, and forms a conspicuous black band around the upper part of Sutherland Mountain. It is vesicular in some parts, and the base of the overlying rhyolite is in most places crowded with basaltic fragments apparently picked up from the scoriaceous upper surface of the flow.

The pink rhyolite forming the seventh of the lithological divisions of the rhyolitic series is exposed at many places throughout the district. It is particularly conspicuous in the west face of Ladd Mountain and in the corresponding slopes of the various hills between Ladd Mountain and Beatty. Its maximum thickness is probably about 60 feet. By reason of its characteristic color and the wide distribution of its exposures it has proved a very useful formation in the study of the faulting undergone by the district, and in Pl. III is distinguished from the other formations in order to illustrate in part the general structure.

Rhyolite No. 8 is the most widespread and extensively exposed of the many rhyolitic formations in the district. It forms the eastern slope of Bullfrog Mountain, the main part of Busch Peak, the summits of Sutherland and Bonanza mountains, and the summit and east slope of Ladd Mountain, and occurs in most of the short ridges between Montgomery Mountain and the hills of pre-Tertiary rocks south of Beatty. This rhyolite is generally dark gray but in many places, as on Busch Peak, has a dull purple tint and, like most of the siliceous members of the volcanic series, exhibits considerable color variation. Quartz phenocrysts are abundant and the matrix being usually rather compact the rhyolite appears to resist erosion better than most of the flows. The lower 60 feet of the formation, whose maximum thickness appears to be about 600 feet, is in some places a volcanic glass.

Rhyolite No. 9 is a gray or nearly white, rather pumiceous flow breccia varying from 25 to 150 feet in thickness. It rests in most places directly upon rhyolite No. 8, but in the hills just north of Rhyolite and on the northeast slope of Ladd Mountain a flow of basalt (feldspar basalt No. 3) intervenes. This flow breccia is eroded with comparative ease and is seldom conspicuous, occurring for the most part in saddles or on the smooth slopes in the eastern half of the district.

Rhyolite No. 10 is a brownish-gray flow breccia containing fragments of rhyolite, basalt, and, rarely, schist. The matrix contains abundant phenocrysts of quartz and clear feldspar. The base of the formation is in many places a rhyolitic glass. The formation is extensively exposed in the eastern half of the district. It occurs in the hills due north of Rhyolite, being in most cases the summit rock, and forms the summits and most of the eastern slopes of Montgomery, Paradise, and Velvet mountains. Its greatest thickness is probably over 400 feet.

Overlying this rhyolite on the northeast side of Velvet, Paradise, and Montgomery mountains, and probably also in the valley west of Rainbow Mountain, is the thickest flow of basaltic rock known in the district. Its maximum thickness may be provisionally estimated at 200 feet. In the field this rock was not supposed to be essentially different from the other basaltic flows in the district. Microscopical study shows, however, that it is a fresh, typical leucite basalt, a type of rock that is not common and that apparently has never been recorded in Nevada. The specimen upon which the present determination rests was taken from an exposure five-eighths of a mile west of Black Peak or about 1½ miles north of Rhyolite. The dark basaltic rock crossed at various points on the road from the Montgomery-Shoshone mine to Beatty is probably of the same type, and a small mass, three-fourths of a mile south of Beatty, is known to be leucite basalt.

The leucite basalt is overlain between Beatty and the Montgomery-Shoshone mine by thin-bedded shales, some of which show worm casts, and by other fine-grained sediments. The total thickness of these beds is probably under 200 feet. They appear to be composed mainly of glassy volcanic detritus and to represent a brief intervolcanic period of sedimentation, possibly in a small basin.

Above these sediments lies rhyolite No. 11, a series of flow breccias which near the base show a very close resemblance to tuffs or mud flows. Indeed, it is not always possible in the field to decide whether a given layer flowed as a molten mass crowded with fragments or as a hot pasty mud. The thickness of this formation is 500 or 600 feet.

Rhyolite No. 11 and the remaining members of the local geological section are all exposed on the west slope of Rainbow Mountain and on Black Peak.

Rhyolites Nos. 12 and 14 are both perlitic glasses, not unlike dark-green bottle glass in appearance. The upper, more prominent band is perhaps 200 feet in thickness. They are separated by rhyolite No. 13, which is also very glassy but is brown instead of green.

Rhyolites 15 and 16 are both nearly white flow breccias, distinguished mainly by the fact that bunches of dense volcanic glass or obsidian are much more abundant in 16 than in 15. No. 16 was the last of the rhyolitic outflows so far as any record has been preserved in this district. It was followed, without any recognized erosion interval, by a thin flow of basalt. which apparently did not entirely

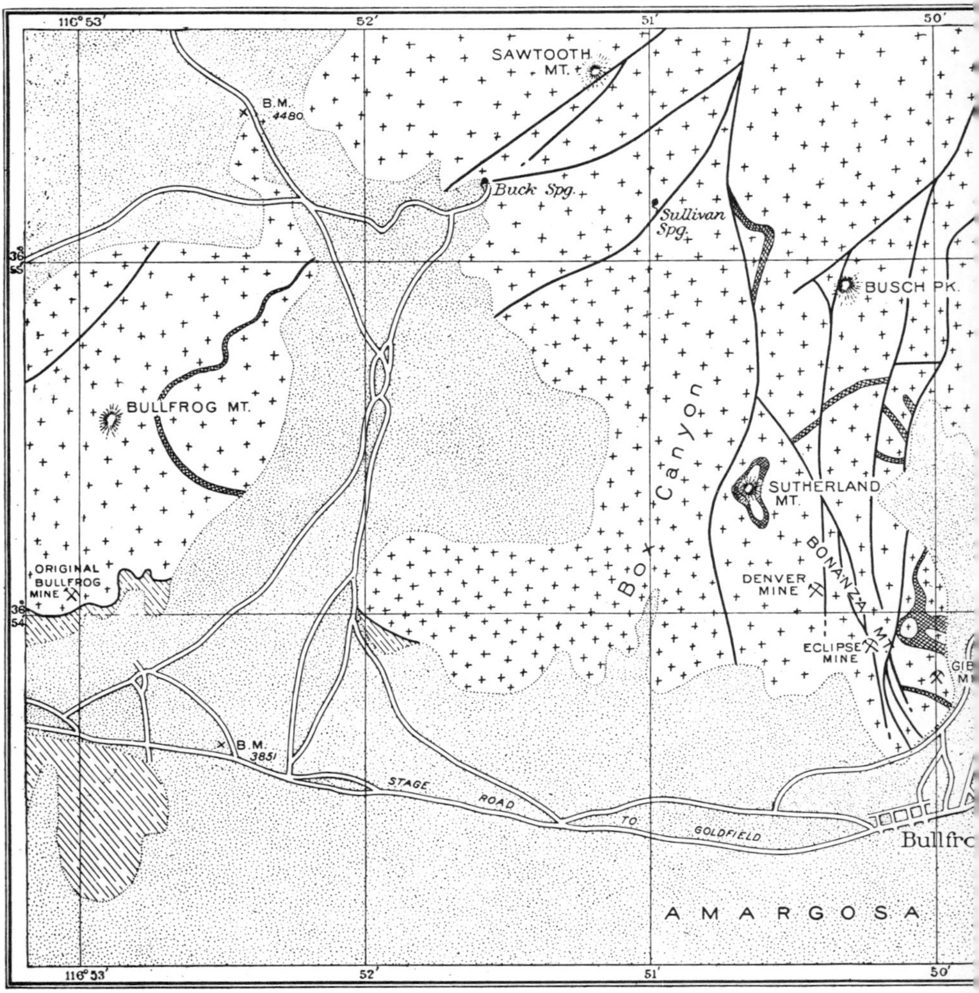

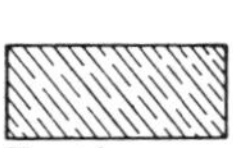

Pre-Tertiary rocks

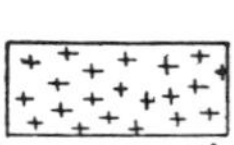

Tertiary volcanic rock
(exclusive of rhyolite No. 7)

PRELIMINARY

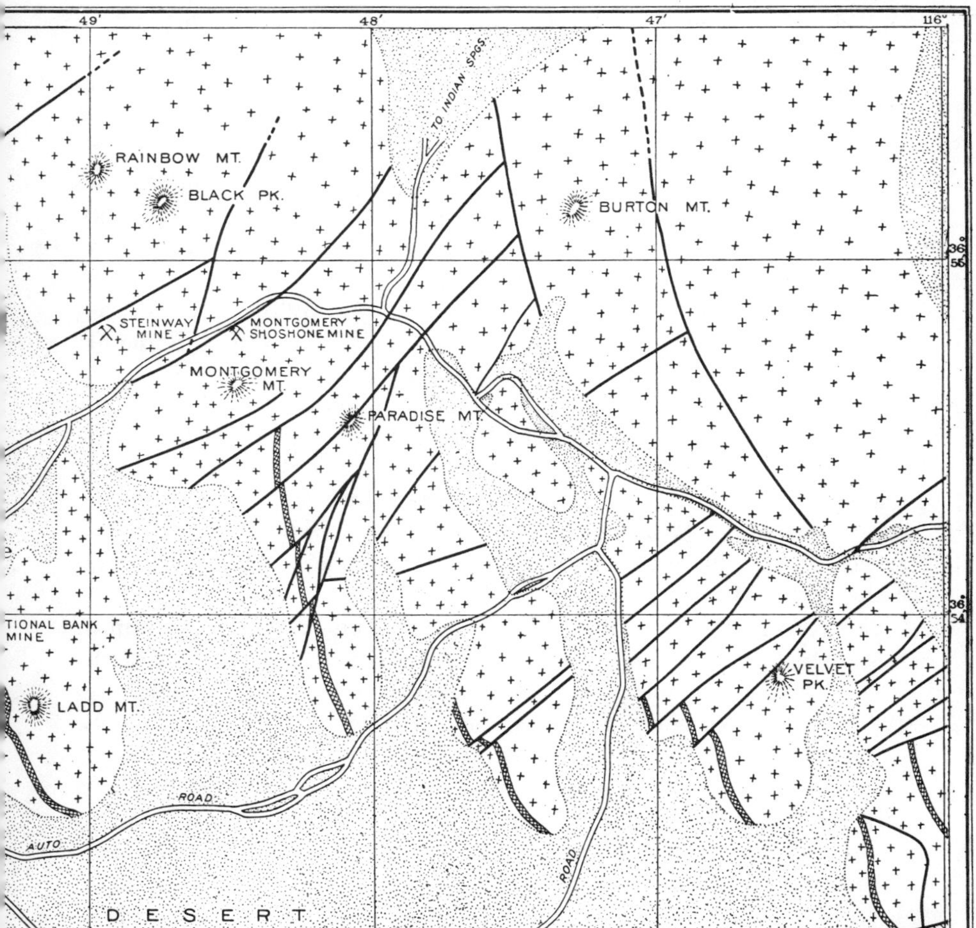

Scale

2 miles

GENDLE

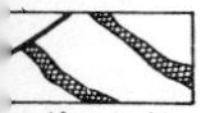

olite No. 7

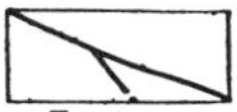

Faults

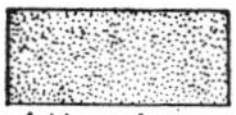

Alluvium

AP OF THE BULLFROG DISTRICT.

cover the rhyolite. Resting partly on the basalt (No. 4 of the feldspar basalts) and partly on rhyolite No. 16 is the dacite flow which caps Black Peak, whose name was evidently suggested by the somber contrast of the dacite to the light and brightly tinted rhyolitic bands of Rainbow Mountain.

Above the dacite, and exposed in the saddle east of the summit of Black Peak, lie some thin beds of sandy tuff, probably not over 20 feet in thickness. These are overlain by a flow of quartz-bearing basalt, the youngest effusive rock in the district. In the district studied the dacite and quartz basalt occur only on the eastern slope of Black Peak. There is reason to think, however, that they have a greater distribution to the north.

No large bodies of intrusive rock occur in the Bullfrog district, but dark, basaltic dikes, usually narrow, are fairly abundant. Some are ordinary olivine basalts, but no microscopic study of the dike rocks has yet been made, and it is possible that leucite or nepheline may occur in some of them. These dikes cut the rhyolitic flows and, although perhaps not all of the same age, they are on the whole so closely related to the faulting as to indicate that they are the latest products of volcanic action preserved in the region. Some of their peculiarities will be described in connection with the structural features of the district in the following section.

The youngest formation in the district is the gravelly alluvium or wash, which, as may be seen from Pl. III, covers a large part of the area. The superficial part of this deposit is composed mainly of angular or imperfectly rounded rock fragments derived from the neighboring hills. Little is known, however, of the character of its basal portion, as the conditions under which the material accumulates are not favorable to the development of deep gorges, such as would furnish natural sections. A shaft has been sunk in the alluvium, about three-fourths of a mile south of the summit of Ladd Mountain, to a reported depth of 330 feet. The dump, which contains some rounded bowlders a foot or more in diameter, indicates that the bottom of the shaft had not reached solid rock when work was stopped.

STRUCTURE.

At the close of the Tertiary rhyolitic eruptions the schists, gneisses, quartzites, and limestones forming the ancient platform upon which the lavas were extravasated had been buried beneath approximately 8,000 feet of nearly horizontal volcanic flows. Over the area of the Bullfrog district these flows appear to have been fairly uniform in thickness. How far beyond this they once extended is not yet known, nor has it been definitely determined whether Bare Mountain, composed of pre-Tertiary rocks, owes its elevation entirely to postvolcanic movement or whether it originally stood above the general surface over which the lavas spread. In the one case the volcanic rocks have been stripped from it by erosion. In the other it may never have been completely buried. The former suggestion is regarded as the more probable.

After the volcanic outflows ceased the rocks were subjected to stresses which were finally relieved by faulting. There is no evidence that the volcanic formations were folded. They appear to have been cut by numerous fissures into many blocks, which were individually tilted by the dislocations. The main fissures trend from north and south to northwest and southeast. They dip to the west in most cases, while the fault blocks are tilted to the east. The faulting is thus normal in character, and one of its most striking results is the repetition, across the whole district, of blocks containing the same assemblages of eastward-dipping formations, as may be seen in Pl. III.

In addition to the dominant faults just referred to there is another set, characterized by nearly northeast-southwest trends. In general the throws of the northeast faults are less than those of the north or northwest faults, but a few northeast faults show profound displacement. Both systems of faults are represented in Pl. III, but the structural importance of the north to northwest faults is somewhat masked by the facts that they are in some cases partly or wholly covered by alluvium and are less numerous than those of the other system. The structure shows that under the band of alluvium between Bullfrog Mountain and the "box canyon" must lie concealed the greatest fault in the district. Another important dislocation must be covered by the alluvium between Ladd and Montgomery mountains.

Most of the faults are nearly vertical, but there are apparently two notable exceptions to this rule. The contact between the rhyolites and the pre-Tertiary rocks in the vicinity of the Original Bullfrog mine dips to the north at an angle of 18° to 20°. It is not well exposed, but seems to be a plane of some disturbance. As the various rhyolitic formations of Bullfrog Mountain dip to the east at considerable angles and are successively cut off at this contact, they must therefore have flowed against a steep slope of Paleozoic rocks or else they have been dropped by a fault. The latter seems on the whole more probable, although it is difficult to understand how normal faulting could take place on a plane so slightly inclined unless the displacement had a large horizontal component.

South of Beatty the relation of the Tertiary volcanic rocks to the pre-Tertiary metamorphic rocks is similar to that just described. The contact is probably a fault, although the evidence for this conclusion is not as satisfactory as could be desired. East of Beatty, however, the rhyolites are clearly faulted down against the limestones and quartzites of Bare Mountain. The fault here dips northward at angles between 50° and 65° and the rocks adjacent to the fissure are more or less crushed.

In general, the rhyolites of the Bullfrog Hills appear to be bounded on the south by a nearly east-west fault or fault zone which has dropped them against a pre-Tertiary metamorphic terrane forming the mass of Bare Mountain and exposed in a few low hills along the northern edge of the Amargosa Desert. The entire downthrown mass has been divided by other faults into numerous small blocks.

A large number of the faults are accompanied by basaltic dikes. In some cases the dike fills the fault fissure and is not itself disturbed. In other cases the faulting is in part later than the intrusion, so that the dike has been fractured or crushed by the movement. A very common feature of these dikes is their lack of continuity, the basalt occurring here and there along a persistent fissure. This peculiarity is probably in part due to movements subsequent to the intrusion, but along many of the fissures the basalt appears to have been intruded only at intervals.

ORE DEPOSITS.

MINING DEVELOPMENT.

The Montgomery-Shoshone, situated about 1½ miles northeast of Rhyolite on the road to Beatty, is the best known mine in the district, on account of both the size and richness of its blocked-out ore body and the legal contests that have been waged for its possession. Negotiations, which afterwards terminated in the sale of the mine, were entered into during the geological study of the district, and were unfortunately considered by the former owners a sufficient reason for withdrawing permission to enter the underground workings. The occurrence of the Montgomery-Shoshone ore has therefore not been carefully studied, although Messrs. Emmons and Garrey, who entered the mine before permission was withdrawn made some preliminary notes, which have been utilized in the present report. The main Montgomery-Shoshone shaft was 150 feet deep at the time of visit, with three levels. The Polaris mine adjoins the Montgomery-Shoshone on the west, the Polaris shaft being about 700 feet southwest of the Montgomery-Shoshone shaft. Northeast of the latter is the Providence shaft, 100 feet deep. The Steinway shaft, about 2,300 feet west of the Montgomery-Shoshone, and the Yankee Girl, in the flat just northeast of Rhyolite, are prospects which at the time of visit were about 200 feet deep. On the south slope of Montgomery Mountain are the Rand and Crystal shafts, 100 feet deep in January, 1906, but not in ore.

On the west slope of Ladd Mountain the most important workings are those of the Bullfrog National Bank Company, a shaft 200 feet deep with two levels, and of the Bullfrog Mining Company, a tunnel that was about 550 feet long on January 19, 1906.

On the steep south slope of Bonanza Mountain is the Gibraltar mine, opened by three tunnels a few hundred feet in length. The upper tunnel (No. 3) is on what is known as No. 4 vein, the one below it (No. 2) is partly a crosscut and partly on the same vein, while the lowest (No. 1) tunnel is a crosscut. No. 1 tunnel enters the east slope of Bonanza only a few feet above the alluvium, No. 2 tunnel is about 85 feet higher, and No. 3 about 60 feet above No. 2.

North of the Gibraltar, and lying mostly on the west slope of the mountain, are the properties of the Eclipse Mining Company and two claims, known as the Tramp No. 1 and Tramp Extension, which are under nearly the same ownership as the Eclipse group. The Eclipse ground is opened by the Eclipse shaft, 125 feet deep; the Hobo incline, 155 feet deep; the Lester shaft, 60 feet deep, and the Tiger tunnel, which was between 300 and 400 feet long at the time of visit. This tunnel enters Bonanza Mountain near its western base and is intended to crosscut the Eclipse and Lester veins below the bottoms of existing shafts. The Eclipse and Tramp groups are still in an early stage of development, no stoping and very little drifting having been done.

About 1,500 feet northwest of the Eclipse shaft is the Denver mine, opened by three tunnels separated by vertical intervals of 70 to 80 feet. A fourth tunnel has been started 100 feet below No. 3. The tunnels in general follow the vein, none of them at the time of the visit being over 300 feet in length.

The principal workings of the Original Bullfrog mine comprise a tunnel with several hundred feet of branching drifts and crosscuts and two shafts, one of which is 140 feet in depth.

The Gold Bar, about 4 miles northwest of Rhyolite, lies outside of the area covered by the detailed map of the Bullfrog district. The main shaft is 150 feet deep, and there are several hundred feet of drifts on two tunnels. Another promising outlying prospect is the Mayflower, situated about 7 miles north of Rhyolite and opened by a shaft which at the time of visit was 100 feet deep. The Happy Hooligan mine, 9 miles a little south of west from the town of Rhyolite and developed by a tunnel and a 40- or 50-foot inclined winze, presents some interesting geological features, which will be referred to later.

There are many other prospects in the district, some of which may ultimately prove more important than one or two of those named. Most of them, however, have been very little developed as yet or are not in ore, and it is not necessary to enumerate them in a brief report of this character.

GENERAL CHARACTER OF THE DEPOSITS.

Attention has already been drawn to the elaborate network of faults that constitutes so important a structural feature of the district. The ore deposits are for the most part nearly vertical mineralized faults or fault zones in rhyolite. Of the many faults shown in Pl. III comparatively few have proved ore-bearing, although it must be remembered that the district is young and that a considerable number of the faults discovered in the course of the geological mapping have not yet attracted the attention of prospectors. The presence or absence of mineralization bears no evident relation to the particular variety of rhyolite constituting the country rock or to the amount of dislocation along the fissures, but appears to be mainly a matter of locality. In certain restricted areas, such as Bonanza Mountain, several fissures will show more or less mineralization, while elsewhere the same and other fissures are barren. As none of the mines has yet gone below the zone of oxidation, direct observation of the original character of the lodes is nowhere possible. All accessible portions of them have been more or less altered by the oxidation of the sulphides, by the solution of some of the gangue constituents, and by later movements along the fissures.

Most of the lodes instead of being simple veins are fissure zones containing numerous stringers of vein material and in most cases showing no definite walls. The principal stringers are parallel with the sides of the lode as a whole, but they are linked by numerous irregular cross veinlets, and similar small stringers extend for varying distances into the country rock. The lodes range in width from a few inches to 10 or even 100 feet. They are in many places accompanied or followed by well-defined regular planes or walls that have resulted from movement along the fissure zone after the original formation of the vein. Such a secondary wall may be seen in the Eclipse and Tramp shafts on the Eclipse vein.

Originally the stringers consisted of quartz and calcite carrying finely disseminated auriferous pyrite. The calcite, possibly associated in some places with other carbonates, varies in abundance. It forms a large part of the Hobo and Louisville veins on Bonanza Mountain, while it is nearly or quite absent from the National Bank vein on Ladd Mountain. The larger stringers, such as those of the Hobo vein, at many places exhibit regular depositional banding or crustification. Many of the stringers are distinct fissure fillings, with a definite contact between them and their rhyolite walls. Cases of transition from vein filling to country rock, due to the silicification of the latter, are by no means lacking, however, although this process has not been carried to anything like the extent observable at Goldfield. Much of the quartz, including some good ore, has a fine granular texture and has evidently formed by the silicification of shattered or crushed rhyolite. Typical vein quartz, such as is characteristic of the gold veins of the Appalachians or Sierra Nevada or such as is found in the pre-Tertiary schists of the Bullfrog district, does not occur in the mines near Rhyolite, with the exception of the Original Bullfrog. The quartz is prevailingly fine grained, often of a porcelainlike texture, and is usually intercrystallized with calcite.

The presence of so much calcite in veins traversing rhyolites, rocks exceptionally low in lime, is not yet fully accounted for. It is known, however, that a considerable part of the pre-Tertiary terrane upon which the volcanic rocks were poured out consists of limestone. There is a suggestion, therefore, that the vein-forming solutions gathered the calcareous constituent of the gangue from these underlying limestones.

In the process of oxidation, which in most of the lodes has been facilitated by movements that have fissured or shattered the original filling, the crystals of pyrite are changed to specks of limonite within which may occasionally be seen particles of native gold. The calcite is partly dissolved and partly changed to fragile cellular pseudomorphs of quartz. In this alteration the silica attacks first the outside of the calcite grain or crystal and works inward along cleavage planes. The final result is a shell of quartz divided by thin quartz septa or partitions, parallel groups of which are in many cases inclined to one another at the characteristic angles of the cleavage rhombohedron of calcite. Between the septa there is usually more or less black or brown earthy material, in part oxide of manganese. The development of earthy hydrous oxide of manganese is very characteristic of the oxidation of the calcitic veins of the district, and large portions of the lodes are made up of soft, dark, manganiferous earth associated with residual masses of the original quartz and calcite and containing vugs and druses of secondary deposits of these two minerals.

The ore bodies of the Montgomery-Shoshone and Original Bullfrog mines are both somewhat exceptional. The principal ore body of the Montgomery-Shoshone is a great, irregular mass of soft kaolinized rhyolite at the junction of a series of nearly north-south fissures with the basalt dike that occupies the Montgomery-Shoshone fault. The Original Bullfrog deposit is a large, nearly horizontal body of coarsely crystalline and chalcedonic quartz which lies in the fault zone between the rhyolite and the Paleozoic limestone.

The association of some of the ore bodies with basaltic dikes raises the question whether the latter have in any way influenced ore deposition. Field study shows that the dikes were intruded prior to the introduction of the gold-bearing solutions and indicates they are related to mineralization only through the fact that they occupied some of the fissures which were subsequently reopened sufficiently to permit ore deposition. There is no reason to suppose that the ores are genetically connected with the basaltic intrusions and it is by no means certain that eruptions of dacite and rhyolite did not continue after the last basaltic dike had been injected.

THE ORES.

All the ore thus far mined or opened up is more or less oxidized and in most cases contains no sulphides. In the Original Bullfrog mine there is a little chalcocite or copper glance, and in some undeveloped veins in the schists south of Beatty there are specks of galena, but the only sulphide thus far found in the other deposits is pyrite. Native gold, alloyed with various proportions of silver, is the only valuable constituent of most of the ores, although cerargyrite or horn silver is fairly abundant in the rich ore of the Montgomery-Shoshone mine and more detailed study may reveal the presence of some of the halogen compounds of silver in the other deposits. The gold is finely divided, and is almost invariably found in the quartz, not in the calcite. Its characteristic association with little limonitic specks, representing oxidized pyrite, has already been referred to. In the rather exceptional Montgomery-Shoshone ore the gold and cerargyrite occur in a soft, impure kaolin formed by alteration of rhyolite. The Original Bullfrog ore consists of quartz carrying free gold, chalcocite, chrysocolla, malachite, and azurite. The quartz occurs in great variety of crystallization and color. Coarsely crystalline radial masses and banded chalcedonic forms are both common, and the colors range from clear white or pale amethyst to translucent yellow, brown, or green. In some parts of the vein calcite is crystallized with the quartz.

The ores, as sacked, range from $100 to $700 a ton, but the deposits on the whole are very much lower in grade than those at Goldfield. It is evident that the future prosperity of the camp must depend upon the successful working of ore much inferior in grade to that sacked for shipment under present conditions. Just what will be the lower limit in value of profitable ore and what is likely to be the general average value of ore for the whole camp under the more favorable conditions expected to follow the completion of a railroad into the district can not yet be determined.

The proportion of silver varies greatly in different mines even when these are close together. In the Denver ore, for example, the silver is said to be almost negligible, whereas the pale gold or electrum from the Gibraltar mine may contain equal parts of gold and silver by weight, although the average proportion of silver is considerably less. In the ore from the Montgomery-Shoshone mine, according to Mr. Ralph I. Johnson, assayer, the average ratio of gold to silver in 160 samples was 1 ounce of gold to 25.44 ounces of silver. This agrees fairly well with the statement made at the mine that about 45 per cent of the value in the $700 ore is in silver.

NOTES ON INDIVIDUAL MINES.

MONTGOMERY-SHOSHONE.

The Montgomery-Shoshone workings are in rhyolite on the southeast side of a basalt dike. The dike occupies a structurally very important fault which dips northwestward. As a result of this dislocation the rocks exposed on Rainbow Mountain and Black Peak, northwest of the fault, which belong to the upper part of the volcanic series, are brought into juxtaposition with the older flows making up Montgomery Mountain, on the southeast side of the fault. The throw of the fault has not yet been carefully calculated, but provisional estimates show that it probably exceeds 2,000 feet. The summits of Black Peak and Montgomery Mountain both lie on the general line of strike of the eastward-dipping flows, but while Black Peak is capped with quartz basalt (see Pl. III), Montgomery Mountain is capped by rhyolite No. 10. At the mine itself the effect of the fault is to bring rhyolite No. 10 on the southeast side of the fault and dike against rhyolite No. 16 on the northwest side. The geological relations of the ore are diagrammatically indicated in fig. 11. The basalt dike is much broken near its walls and a considerable part of the faulting is later than the intrusion. How much of the total throw was effected before the injection of the dike can not be determined. In the Providence mine the basalt has been crosscut and a drift has been run along its hanging wall, which is a remarkably regular surface of rhyolite polished by the movement of faulting. All of the basalt is fractured and disturbed, and near the hanging wall the squeezed fragments have curved, glossy surfaces, recalling those common in many large bodies of serpentine.

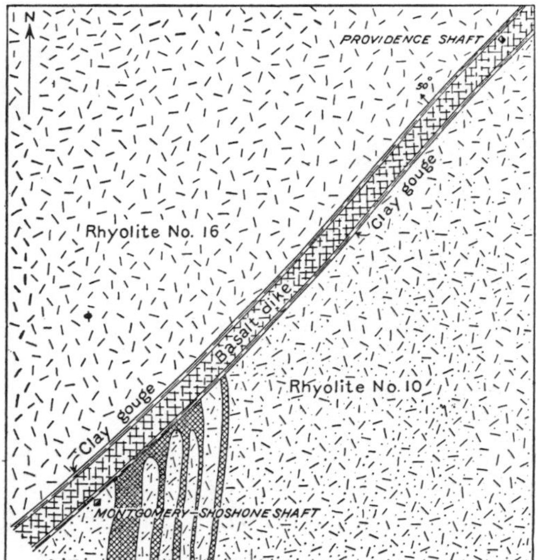

FIG. 11.—Diagrammatic plan showing mode of occurrence of ore in the Montgomery-Shoshone mine.

The fissures that have determined the ore deposition in the Montgomery-Shoshone mine strike, in general, a little east of north. They appear to terminate at the basalt, since no continuation of them has been found on the northwest side of the dike. Along these fissures the rhyolite has been altered to soft masses of kaolin separated, as shown in fig. 11, by ribs of comparatively unaltered rock. Near the basalt the kaolinized zones widen and, in some cases, coalesce. Here occur the largest and richest ore bodies, the principal one, said to average $700 a ton, being approximately 60 feet in horizontal diameter. Along the line of the fissures southward, away from the basalt, the ore becomes of lower grade and there is less kaolinization, the soft material changing gradually into crushed and somewhat silicified rhyolite.

The determination of the relative age of the north-south fissures and of their mineralization, on the one hand, and of the basalt dike and the faulting on the other, is a problem of some practical importance. Are the fissures older than the dike and have their northern continuations been displaced by the great northeast-southwest fault? Or are they younger than the dike and have they no continuation beyond it? Unfortunately the careful study that might have afforded some answer to these questions was rendered impossible at the time of visit by the attitude of the company then in control of the mine. It is thought, however, that the mineralization and probably the fissuring along which ore deposition took place are younger than the dike. The absence of ore from the basalt itself and from the main fault, except where the north-south fissures meet it, offers no serious objection to this view, whereas the coalescence of the ore bodies and the increase in their size and richness near the dike are distinctly in its favor.

NATIONAL BANK.

As exposed at the surface and on the 100-foot level, the National Bank vein is a rather indefinite zone of fissuring in rhyolite No. 5. It strikes about N. 15° E. and dips west at 60°. The hanging wall is a basalt dike with an average width of 3 to 4 feet. The lode has no recognizable foot wall and consists of somewhat silicified rhyolite cut by many small and inconspicuous veinlets of quartz. The gold is not confined to the veinlets, but occurs also in the rhyolite, particularly where the latter shows little limonitic specks representing oxidized pyrite crystals.

The ore is bunchy and good assays are sometimes obtained from the rhyolite at a distance of 30 or 40 feet from what is considered the main lode. The best assays obtained prior to the date of visit were about $230 a ton, mostly in gold. In the lower grade ore the silver may exceed the gold, in one case there being about 5 ounces of silver to 3 of gold.

Considerable work has been done on the 200-foot level, mainly on a narrow seam which dips eastward at an angle of 45° to 50°, and which is probably the contact between two rhyolitic flows. The dip of the National Bank lode from the surface to the 100-foot level would, if continued to the 200-foot level, carry the vein from 50 to 60 feet west of the shaft, or from 15 to 20 feet west of any crosscut in existence on this level at the time of visit.

GIBRALTAR.

There are six veins recognized in the Gibraltar property. Their strikes range from N. 15° W. to N. 20° E. and they dip to the west. The veins are numbered from east to west. No. 4, on which most work has been done, strikes about N. 10° W. and dips west at angles varying from 45° to 55°. No. 2 vein is the most nearly vertical, its dip at the surface being about 80°. The croppings of the veins are spaced at various distances apart, ranging from 50 to 150 feet.

The general country rock of the mine is a rhyolitic flow breccia containing occasional inclusions of basalt up to 3 feet in diameter. It has been correlated with rhyolite No. 6, although the many faults in Bonanza Mountain render the interpretation of the structure less satisfactory than in other parts of the district. The rock is considerably altered and contains much secondary quartz.

The veins are small and occupy fault fissures of slight throw. No. 4 vein, which has been opened by two tunnels and has supplied all the ore so far found, is in most places less than a foot wide, although local widths of 3 to 4 feet occur. It consists of quartz and a soft, dark, earthy material, which is in part oxide of manganese. A cavernous and platy structure, due to the solution of carbonates from an originally banded vein, is common. The best ore lies near the foot wall and usually includes a little of the rhyolite. The gold, which contains so much silver as to constitute electrum, occurs in the mode characteristic of the district, usually in little limonitic specks in quartz or silicified rhyolite. These rusty specks in rare instances show small residual grains of pyrite.

The veins in general cut two older structures in the rhyolite. One of these is a set of planes dipping east at angles ranging from 30° to 35° and parallel to the flow bands in the rhyolite. These may be partings between separate flows, or possibly cracks formed by the cooling of the rhyolite. The other structure is a fairly regular sheeting striking N. 30° W. and dipping southwest at an angle of 60°. When the lode fissures were formed they appear to have occasionally followed one or the other of these earlier fissures for short distances, so that the lodes are unusually crooked, as may be well seen in No. 3 tunnel on No. 4 vein.

ECLIPSE AND TRAMP.

The principal veins in these properties are the Hobo, Lester, and Eclipse. Their general relations may best be understood from fig. 12, for which no pretense of accuracy is made. The Hobo shaft, situated on the crest of Bonanza Mountain, is an incline on the Hobo vein, which at this point appears to strike a little west of north and dips west at angles ranging from 50° to 55°. Just north of the shaft the vein is cut by a fault, marked by a smooth, regular slip plane which strikes N. 10° E. and dips west at nearly the same angle as the vein. Too little work has been done to determine how this fault affects the Hobo vein, but the latter shows much disturbance in its vicinity, and is probably cut off obliquely north of the shaft. The same fault can be traced northward along the east slope of Sutherland Mountain and over the slopes east of Busch Peak. Little masses of basalt occur here and there along its course, and it probably represents an old fissure which, after being filled by a dike, became again a zone of faulting. Thus the Hobo vein may never have extended north of the point where it now terminates against a plane that, while it shows at least some movement since the vein was formed, may also have been a barrier to solutions when the ore was being deposited. The conditions, in other words, are similar to those at the Montgomery-Shoshone mine.

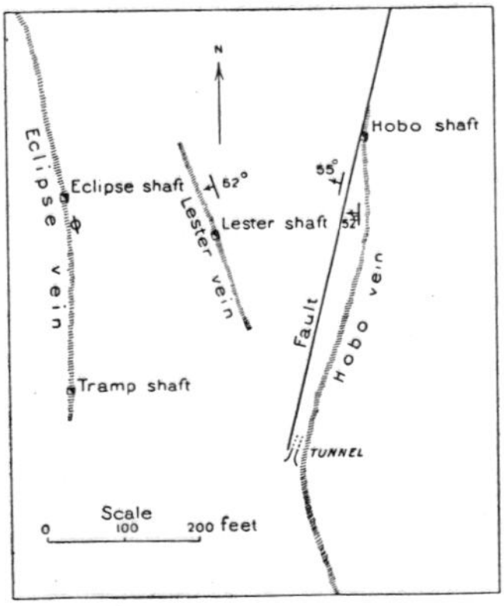

FIG. 12.—Sketch plan of the veins of the Eclipse and Tramp groups.

South of the Hobo shaft the vein can be followed for a short distance along the ridge and then it turns down the west slope of the mountain toward the mouth of the tunnel shown in fig. 12. On the ground this curve is exaggerated by the relation of the dip of the vein to the slope of the hill. Near the portal of the tunnel the fault and the vein again come together. South of the tunnel a very large vein continues along the west slope of Bonanza Mountain and down the south slope toward Bullfrog. This vein, as may be seen from fig. 12, has the general course of the Lester vein, but it appears to be continuous with the Hobo vein.

As exposed in the Hobo shaft the vein consists of a broken mass of quartz, calcite, and earthy oxide of manganese, the last being a residue from the solution of a manganiferous carbonate and the quartz and calcite being partly secondary. The vein is small near the surface, but becomes larger and more disturbed with depth. The best ore, which affords assays up to $400 a ton, is a fine-grained, opaque, white quartz in which free gold occurs in small limonitic specks. South of the tunnel the vein is less disturbed, and in places shows from 25 to 50 feet of fissured and shattered rhyolite full of banded veinlets of quartz and calcite. This part of the lode is apparently of low grade.

The Lester vein is a zone of fissured and veined rhyolite a foot or two wide, with a regular hanging wall. It strikes N. 18° W. and dips west at 52°. An inclined shaft 60 feet deep has been sunk on the vein, but no shipping ore has yet been found.

The Eclipse vein coincides with an important fault which passes through the saddle north of Sutherland Mountain. The general course of the fissure is thus nearly north-northwest, but between the Eclipse and Tramp shafts the vein runs north and south. At the Eclipse shaft the lode, which is practically vertical, has a maximum width of 12 feet. Samples across this width are said to give assays of $34 a ton. The vein is similar in general character to the Hobo and has no very definite walls. The lode itself, however, is traversed by a smooth and regular plane of faulting.

DENVER.

The Denver vein occupies part of a fault or fault zone that can be traced northward over Busch Peak. The fissure that has hitherto furnished the ore strikes from N. 5° to 10° E. and dips east at angles ranging from 65° to 70°. This, however, is merely one of several branching fissures, some of which strike nearly northwest. All the fissures traverse rhyolite No. 5, which is much cracked and somewhat silicified in their vicinity. The Denver lode is generally similar in character to the Hobo and the Eclipse. The streak of shipping ore varies in width up to a maximum of 18 inches.

ORIGINAL BULLFROG.

The Original Bullfrog lode outcrops more conspicuously than the other deposits in the district. It is a huge mass of nearly solid quartz, which, as a whole, dips north at 18° to 20° and is at least 60 feet thick. This quartz rests in most places upon a much sheared shaly material, greenish or reddish in color, which was probably originally a glassy rhyolitic flow. It is now so soft and altered, however, that its original character must remain somewhat in doubt. This material is not of great thickness and in some places the quartz rests directly upon the Paleozoic limestone. The lode has no definite hanging wall, but is overlain by rhyolite No. 2, which is fissured and contains stringers of quartz for some distance above the mass of the lode. The deposit represents a mass of rhyolite that has been greatly fissured and shattered. The fissures have been filled with quartz and with minor amounts of calcite and ore minerals, and to a considerable extent the shattered rhyolite has been completely silicified.

Some bunches of rich ore have been found, but the mass as a whole is of very low grade. The fissuring that provided opportunity for the deposition of so much silica was probably caused by movement along the fault which has here brought the rhyolites against the pre-Tertiary rocks. A deposit having some analogous features with the Original Bullfrog and known as the Wildcat occurs about a mile southeast of

Beatty on the line of the fault between the rhyolitic series and the older rocks. Here, however, it is the limestone that is irregularly silicified and stained with malachite. The ore, which is apparently only a small bunch, carries silver and a little copper. It is noteworthy that the only occurrences of copper minerals noted in the district should be situated so far apart on the same zone of faulting between the rhyolites and the pre-Tertiary rocks.

GOLD BAR.

The Gold Bar lode is a zone of irregularly fissured and brecciated rhyolite fully 100 feet wide. The hanging wall is fairly regular and persistent and is apparently due to a slip later than the formation of the lode. No foot wall has yet been found. The general strike of the hanging wall is N. 50° E. and the dip is northwest at 65°. Most of the stringers making up the lode are filled with quartz, although calcite and siliceous pseudomorphs after calcite occur, particularly near the hanging wall.

The best ore occurs near the hanging wall and is said to average about $30 a ton, but occasional small bunches are very much higher in grade.

HAPPY HOOLIGAN.

At the Happy Hooligan mine, limestone, probably Paleozoic, is overlain by a flow of basalt estimated to be 40 to 50 feet in thickness. The basalt is in turn overlain by a much thicker flow of rhyolite. The surface of the limestone is hummocky, slopes of 20° to 30° being not uncommon. A little oxidized sedimentary material occurs at some points between the basalt and the limestone.

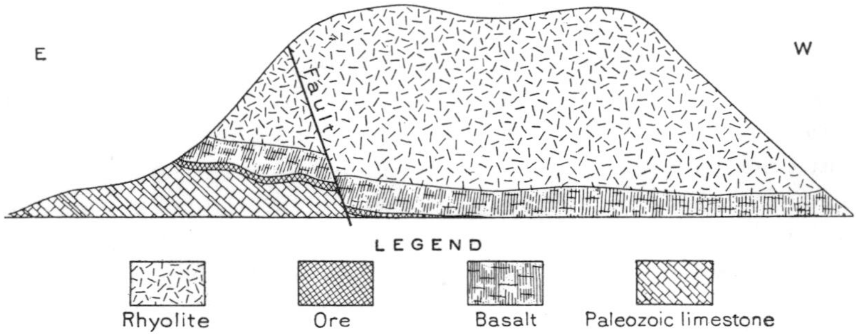

FIG. 13.—Diagrammatic section through the Happy Hooligan mine.

The ore forms a thin, soft undulating blanket deposit between the basalt and the limestone and consists of a waxy white and brown material (probably kaolin, although it has not yet been chemically examined) carrying free gold. Some of the ore appears to be altered basalt and some may represent the alteration of tuffaceous material that lay on the limestone when the basalt was poured out. The ore is rarely more than a few inches in thickness.

The deposit outcrops along the east slope of a small hill and has been followed westward by a tunnel to a point where the ore is cut off by a fault. The fault plane dips west at an angle of 60°. A winze on the fault was 20 feet deep at the time of visit and showed a basalt hanging and limestone foot wall. The structural features of the deposit are diagrammatically shown in fig. 13.

The source of the gold that has been locally concentrated along the contact between the Tertiary volcanic rocks and Paleozoic limestone is not known. The gold is flaky and has none of the characteristics of placer gold. The character of the matrix, moreover, suggests deposition from solutions capable of effecting considerable alteration in the rocks traversed by them.

No ore has yet been shipped, and future developments depend largely upon the conditions that may be disclosed by the winze now being put down on the fault.

SEARCHLIGHT AND ELDORADO DISTRICTS.

INTRODUCTION.

The Searchlight and Eldorado districts were visited early in February, 1906, partly in response to a petition from all of the principal mine owners or mine managers to the Secretary of the Interior asking for detailed geological work in the vicinity. The main object of the visit was to procure such general information as would enable future topographic and geological work to be laid out to the best advantage. In all about four days were spent looking over the Searchlight district and one in Eldorado Canyon. The following notes are necessarily fragmentary, but may be worth recording as a preliminary sketch of a region concerning which very little has been published and in which active development is now in progress.

SITUATION.

The Searchlight district is in the extreme southern point of Nevada in Lincoln County, 12 miles west of the Colorado River and about 50 miles north of Needles. Two stage lines, approximately 25 miles in length, connect the town of Searchlight with the nearest railway stations in California, which are Nipton, on the San Pedro, Los Angeles and Salt Lake Railroad, 56 miles south of Las Vegas, Nev., and Manvel, on the Ivanpah branch of the Atchison, Topeka and Santa Fe system, 30 miles north of Blake on the main line. Railways are projected from both Nipton and Manvel into Searchlight.

Eldorado Canyon heads about 22 miles north of Searchlight and is 10 miles in length. It extends east and west, embouching on the Colorado River. There are no regular means of communication with the Eldorado Canyon district, which, however, is connected by a good road with Searchlight. A weekly mail service was formerly maintained between Chloride, the terminus of the Arizona and Utah Railroad, which connects with the Santa Fe system at Kingman, Ariz., but whether a stage was running over this route early in 1906 could not be ascertained.

No recent surveys have been made of this part of Nevada, and existing maps are all more or less obsolete and inaccurate. On most of them the town of Searchlight is either not shown or is only approximately located. The general positions of Searchlight and Eldorado Canyon are indicated, as nearly as can be determined, in the accompanying index map (fig. 1, p. 9), and the relations of the two districts to each other and to the Colorado River are shown in the sketch map forming Pl. IV.

HISTORY.

Mining in Eldorado Canyon is said to date from about the year 1857, and the Techatticup (an Indian name, said to signify "plenty for all"), one of the principal mines, is reported to have been in intermittent operation since 1863. In 1883 the same mine was worked with 23 men, the ore being treated in a 15-stamp mill. The district bears evidence of having been extensively worked for many years past, large bodies of ore having been stoped from the Quaker City, Mocking Bird, Wall Street, and Techatticup mines. Latterly, however, the only work on this group, now under one ownership, has been confined to prospecting the Techatticup lode below the old stopes.

It is a little surprising that a district once alive with activity should have attracted so little outside notice. This, however, is partly accounted for by the overshadowing predominance of the Comstock, Eureka, Ely, and other districts noted in the early history of mining in Nevada and by the isolation of Eldorado Canyon. The rich ore shipped from the canyon in early days was taken down the Colorado by boat to Needles, Yuma, or the Gulf of California.

The Searchlight district is of comparatively recent development. Ore was discovered early in 1898 on the Searchlight claim, now part of the Duplex mine, and some of the richest ore found in the district came from the stopes above the 200-foot level on this lode. The principal mine at present, the Quartette, was opened in the same year, and in 1902 a 20-stamp mill was erected on the Colorado River. This was shortly afterwards connected with the mine by 15 miles of narrow-gage railway. In June, 1903, after the mill had been running about six months, the miners in the district struck, and after three months' idleness the Quartette mine was reopened with nonunion men. About this time water was encountered in the lower levels and this led to the construction of a new 20-stamp mill at the mine and the abandonment of the project for milling at the river.

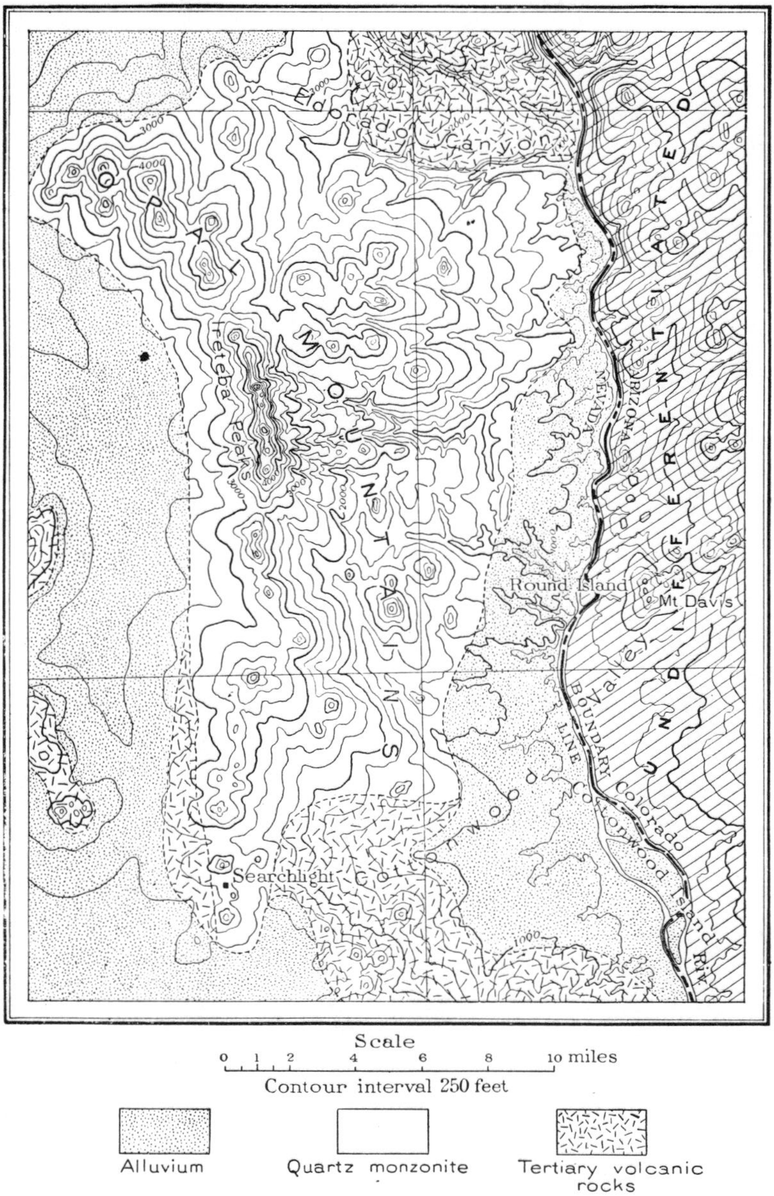

GEOLOGIC SKETCH MAP OF THE SEARCHLIGHT AND ELDORADO DISTRICTS.

PRODUCTION.

The total output of the Eldorado district is not known and is difficult to estimate. It may be anywhere between $2,000,000 and $5,000,000. The Techatticup mine is said to have produced $1,700,000, but this estimate could not be verified. The production of the Searchlight district to the end of the year 1905 probably lies between $1,750,000 and $2,000,000. Of the total sum the Quartette mine has produced approximately $1,250,000.

TOPOGRAPHY.

The Eldorado and Searchlight districts both lie in a range of hills known as the Opal Mountains and locally marking the edge of the diversified plateau that in this region borders the valley of the Colorado on the west. The general altitude of this plateau as defined by its broad desert plains is from 2,000 to 3,000 feet above sea, or from 1,000 to 2,000 feet above the river. The town of Searchlight is approximately 2,200 feet above the Colorado, which here flows in a valley whose width varies from 10 to 15 miles, although very little of this is occupied by actual flood plain. At Searchlight the hills rise to a maximum height of about 1,000 feet above the alluvial plains and are not particularly rugged. At Eldorado Canyon the general slope from the plateau edge to the Colorado River is rather steep and is deeply scored by erosion, so that contrasts in relief are more marked than at Searchlight.

GEOLOGY.

In the very brief time spent in these districts only the most obvious and general relations of the rocks could be ascertained. The accompanying geological sketch map is merely a rough diagrammatic representation of the apparent distribution of the rocks, and the contacts as sketched may be a mile or more from their actual positions. Such a sketch, valueless as to details, will, however, if its limitations are clearly understood, assist the reader in understanding the broad features of the geological structure.

The essential facts of this structure are the existence of a north-south belt of quartz monzonite (locally known as granite), gneiss, and schist, which forms the main mass of the Opal Mountains and is flanked here and there by areas of Tertiary volcanic rocks. The schists and gneisses are the oldest rocks in the region and may be pre-Cambrian. They are much disturbed and are cut by the quartz monzonite as well as by dikes of aplite, pegmatite, and andesitic porphyries. Near Searchlight these older rocks are represented by biotitic gneiss forming part of the ridge that extends south of town and east of the Quartette mine. North of the town for a distance of 15 miles the Opal Mountains appear to consist mainly of quartz monzonite, with possibly some granite. At the Nob Hill camp, however, situated on the crest of the ridge, 4 or 5 miles south of Eldorado Canyon, the prevailing rocks are micaceous and hornblendic schists and biotitic gneisses. Between Nob Hill and the head of Eldorado Canyon are rounded hills of gneiss and schist, cut by irregular dikes of pegmatite, aplite, fine-grained quartz monzonite, and various andesitic porphyries. These hills here mark the edge of the general upland, which descends in a rugged canyoned slope to the river. So far as could be seen in the foreground of the grand panorama spread out eastward from Nob Hill the schists and gneisses extend to the Colorado on the south side of Eldorado Canyon. The upper part of Eldorado Canyon, particularly the north side, is chiefly in quartz monzonite.

The "granite" in Eldorado Canyon and near Searchlight is uniformly a fine-grained gray rock, showing to the naked eye abundant biotite, some hornblende, and apparently considerable quantities of plagioclase as well as orthoclase. Titanite is a noticeable accessory mineral. The quartz is very inconspicuous, and the general appearance of the rock suggests a fine-grained monzonite rather than a true granite.

This suggestion is corroborated by the microscopic study of thin sections. These show that plagioclase and orthoclase are about equally abundant. Quartz is a more important constituent than would be supposed from a cursory examination, it filling many of the interstices between the partly idiomorphic feldspars. The principal dark constituent is biotite, which, however, is in most places accompanied by a little green hornblende and a colorless monoclinic pyroxene. Titanite, apatite, and magnetite are all fairly abundant accessory minerals. The plagioclase, although probably not all of the same composition, corresponds in general to andesine. An estimate of the volumetric proportions of the various minerals in a specimen from Eldorado Canyon afforded a basis for a calculation of the following approximate mineralogical composition of the rock by weight:

Approximate mineralogical composition of quartz monzonite.

Andesine	34.7	Augite	2.3
Orthoclase	34.0	Titanite	.8
Quartz	12.6	Apatite	.5
Biotite	6.9		
Magnetite	4.4	Total	100.0
Hornblende	3.8		

The rock is thus, in all probability, a quartz monzonite, intermediate in composition between a true granite and a quartz diorite.

The quartz monzonite, like the gneiss and schist, is supposed to be a part of the old crystalline floor through which the Tertiary eruptives broke. Definite proof of this, however, was not obtained in the course of the present reconnaissance, and the possibility of its being a Tertiary intrusive mass is still open to consideration.

Bordering the quartz monzonite and schist belt on the east, near Searchlight, and lying between these rocks and the river, is a zone of volcanic rocks several miles in width. East of Searchlight the volcanic series is several hundred feet thick and consists of flows of hornblendic andesite, beds of andesitic breccia, and white and brown tuffaceous sandstones and some minor flows of basalt. The whole series of effusive lavas and tuffaceous beds dips to the west, the maximum dip observed being 55°. This general westerly dip is maintained close up to the quartz monzonite (fig. 14), and it is difficult to avoid the conclusion that the contact is a fault. No definite fault plane, however, could be discovered in the brief time devoted to this area, and along the general line where the fault might be expected there are usually irregular intrusive bodies of a biotitic andesite porphyry, which, on account of its conspicuous white phenocrysts of feldspar, is

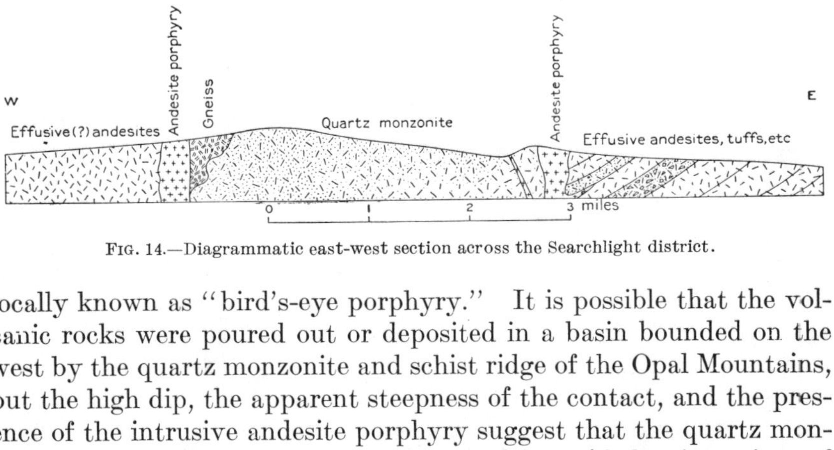

FIG. 14.—Diagrammatic east-west section across the Searchlight district.

locally known as "bird's-eye porphyry." It is possible that the volcanic rocks were poured out or deposited in a basin bounded on the west by the quartz monzonite and schist ridge of the Opal Mountains, but the high dip, the apparent steepness of the contact, and the presence of the intrusive andesite porphyry suggest that the quartz monzonite has been relatively upthrust by faulting and that intrusions of porphyry took place along the zone of dislocation.

On the west side of the quartz monzonite, near Searchlight, the relations of the rocks appear to be generally similar to those just described, although they are even less clearly shown. West of the monzonite and gneiss are low slopes of andesite passing westward beneath a cover of alluvium and traversed near the granite by numerous dikes of andesite porphyry. The same porphyry has also invaded the gneiss south of Searchlight in a multitude of nearly parallel dikes, in a way that will be more fully described in connection with the Quartette mine. Whether the hornblendic andesites west of Searchlight are flows or intrusive masses is a question that only detailed study can satisfactorily answer. They are probably flows.

On the north side of Eldorado Canyon volcanic rocks occur in great thickness between the abandoned town of Nelson, about 7 miles from the river, and the mouth of the canyon. The lowest member of the series is a tremendous flow of purplish-red, oxidized basalt, fully 1,000 feet in thickness and vesicular throughout. This rock forms picturesque cavernous cliffs just north of Nelson. It is overlain by a nearly equal thickness of tawny and buff rhyolite flow breccia. These flows seem to be entirely later than the mineralization in the upper part of the cañyon and may be younger than the andesites near Searchlight. They evidently filled a valley in the older rocks, since the rhyolite between Nelson and the Techatticup mine (a mile or two east of the town site) rests directly upon an uneven surface of granite. These flows, unlike those east of Searchlight, dip toward the river.

MINES OF THE SEARCHLIGHT DISTRICT.

DISTRIBUTION.

The principal mines and prospects in the Searchlight district are in a north-south belt, about 4 miles long and three-fourths of a mile wide, which lies just west of the town. On the east side of the ridge, between Searchlight and the river, are a number of scattered prospects, none of which has yet produced important quantities of ore.

GENERAL CHARACTER OF THE DEPOSITS.

With the exception of the Southern Nevada, the mines of the belt west of Searchlight that were visited (and probably also those not entered) are on fissure zones striking approximately N. 65° W. and dipping southwest at various angles. All are within half a mile of the general contact between the quartz monzonite and gneiss on the one hand and the Tertiary eruptives on the other. In some mines, such as the Quartette and possibly the Duplex, the lode crosses the contact and extends into the area of the older rocks.

The lodes contain very little solid quartz and do not outcrop prominently. Toward the west they either pinch out or pass beneath the alluvium that covers the lower slopes of the hills. They are essentially mineralized fault zones in which the original character of the mineralization has been obscured by repeated movement and by oxidation. The lode material is generally a soft mass of shattered or crushed country rock colored by chrysocolla and oxides of iron and carrying free gold as its valuable constituent. Quartz is common in vugs, druses, and veinlets. Cerussite is a characteristic constituent of the Quartette ore and is associated with wulfenite, cuprodescloizite, and leadhillite. Other lode minerals are specular hematite, malachite, azurite, and calcite. The only sulphides noted are chalcocite, galena, and pyrite, which are found in small quantities in the Quartette mine.

The material thus far stoped is almost wholly free-milling gold-silver ore. It is evident, however, that below the zone of oxidation ores may be expected that will not yield to direct amalgamation.

NOTES ON INDIVIDUAL MINES.

QUARTETTE MINE.

The Quartette Mining Company owns three groups of claims, one comprising the Golden Treasure and Copper King claims, situated about three-fourths of a mile southwest of Searchlight, another known as the Boston group, which is about 1¼ miles northeast of town, and the third consisting of a group of comparatively undeveloped claims near Fourth of July Mountain, about 3 miles east of Searchlight. Operations are at present confined to the Golden Treasure and Copper King claims, in what is generally known as the Quartette mine.

The Quartette lode occupies a strong fissure zone striking from N. 65° W. to N. 68° W. The dip is to the south and is rather variable, ranging from 20° to 70°. The average is probably about 50°. The mine is opened by an inclined shaft on the lode, the shaft pitching somewhat to the west. The total depth is 850 feet on the incline or 600 feet vertically. There are 9 levels, rather unevenly spaced. The deposit on the whole is regular and easily followed, so that there has been little exploration of the country rock in the hanging and foot walls. Almost all of the development thus far has been to the east of the shaft, in which direction the outcrop of the lode can readily be followed over the ridge toward the Rambler mine. The vein has been opened at several points, particularly on the Copper King claim near the crest of the ridge, and in most places is easily recognized by the presence of quartz, chrysocolla, and hematite. West of the shaft the vein can be followed only a short distance.

The general geological relations of the deposit are shown in Pl. V. Prof. T. A. Jaggar, of the Massachusetts Institute of Technology and Harvard University, spent a few weeks in the district in 1903 and prepared a report on the geology of the Quartette mine. The geological boundaries shown in Pl. V were taken from a map accompanying this report. The contours, however, are sketched from memory. The contact between the andesite and the older rocks, here represented by biotitic gneiss with some aplite dikes, is marked by a ravine just east of the mine buildings. East of this ravine rises the main ridge south of Searchl ght, composed of gneiss which is cut by a multitude of nearly parallel dikes of andesite porphyry. These dikes have a general strike of N. 65° W. and follow the foliation of the gneiss. Most of them appear to terminate at the contact between the gneiss and andesite, but some continue into the latter rock, which contains also some less regular intrusions of the same porphyry. The rocks in the vicinity of the lode, particularly the andesite, show considerable alteration, which, joined with rather poor exposures and the great number of the closely spaced dikes, renders the working out of the structure a matter of some difficulty. The underground workings, unfortunately, do not promise much assistance in the elucidation of the local geology.

The lode varies in width, the maximum measurement being 50 feet, and is in general a soft oxidized mass consisting of shattered and mineralized gneiss, andesite, and andesite porphyry. The hanging wall is usually defined by a seam of soft, moist clay, but the rock is much broken and is seamed with veinlets of quartz and calcite for an unknown distance back of this wall. In some places there is a similar gouge along the foot wall, but as a rule this wall is not well defined. Subsidiary slips, similar to those along the foot and hanging walls, are often found in what is regarded as the body of the vein. Owing to the general oxidation it is impossible to get a satisfactory conception of the original character of the vein. In its present form it is a complex structure modified by brecciation and by more or less migration and redeposition of its mineral constituents. The country rock near the lode is in most places much crushed and altered, and it is doubtful whether, without more numerous crosscuts, the complex relations of andesite, andesite porphyry, and gneiss can be satisfactorily worked out. While the lode dips to the south, the dikes as a rule dip steeply to the north. The result is that abrupt changes in the character of the country rock are very common in the mine. No relation has so far been discovered between kind of country rock and value of ore.

As no stope maps are kept, it is not practicable to ascertain the exact shape of the pay shoots. The first discovery of ore was at the surface, about 100 yards west of the present shaft, where a rather irregular bunch was extracted which apparently had no connection with ore afterwards found in the mine. Ore occurs at the surface also at the Copper King tunnel near the crest of the main ridge east of the shaft. The main workings of the mine, however, are not yet under this point. In the mine proper there was practically no ore on the first level and very little on the second. Between the second and third levels, however, a shoot of rich ore was found east of the shaft, which pinched out 2 or 3 feet above the third level. At the shaft on the third level a body of good ore was stoped to a width of 30 feet. The shoot was short, however, and the third level as a whole has developed little ore. Below the third level the stopes are fairly continuous and wide down to the seventh level. The ore body has been

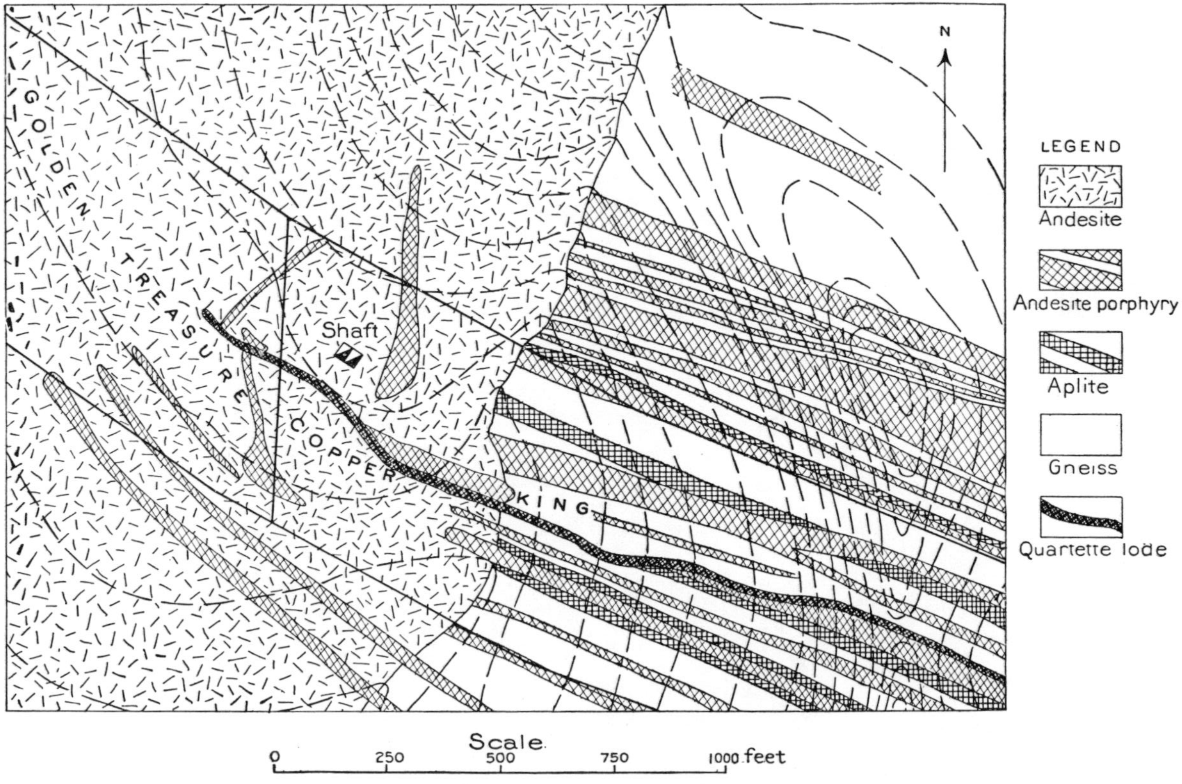

GEOLOGIC SKETCH MAP OF THE VICINITY OF THE QUARTETTE MINE.

particularly productive between the fifth and sixth levels, the latter being regarded as the best level in the mine. In general the main ore shoot is about 300 feet long and up to 50 feet in width. It apparently pitches to the east. At the time of visit no stoping had been done on the eighth and ninth levels. The former had been driven about 300 feet to the east, but apparently had not reached the ore shoot, which on the sixth level is at the shaft. A good shoot of ore, however, supposed to pitch to the west, was being opened on the west drift of the eighth level. The ninth level had only short drifts and was not yet in ore. Where the ore body pinches, the lode may contract to a thin seam of gouge between walls of somewhat shattered country rock, or it may continue as a broad zone of crushed and altered rock having the general appearance of ore, but without value.

The ore is ordinarily a soft, shattered mass of altered country rock intimately mingled with the various ore minerals. It is generally oxidized, but a little chalcocite is said to be occasionally found and some residual kernels of galena were seen in the west drift on the eighth level. Good ore is nearly always associated with chrysocolla or with earthy cuprite, and in some places carries from 10 to 12 per cent of copper. Small cupriferous bunches are sorted and shipped. Where galena or, more commonly, cerussite appears the ore carries a larger proportion of silver than where copper is the dominant base metal. Wulfenite, the molybdate of lead, in characteristic square, tabular, orange-colored crystals, is very common throughout the mine and does not necessarily indicate ore. It is apparently one of the later products of oxidation and occurs in little vugs and open fissures implanted on the other minerals. Hematite is abundant, but, like the wulfenite, does not seem to be particularly characteristic of good ore. Quartz is widely distributed through the lode but never forms large or solid masses. Its most typical occurrence is in small vugs and veinlets. A little malachite and azurite frequently accompany the chrysocolla, and cerussite (carbonate of lead) is fairly abundant. Cuprodescloizite (a basic vanadate of lead, zinc, and copper) and leadhillite (a sulphato-carbonate of lead) were determined by Prof. Charles Palache, of Harvard University, in specimens collected by Professor Jaggar. Vanadinite (vanadate of lead) and native copper are reported as occasional and inconspicuous constituents of the ore. Visible specks of gold are not uncommon in the rich ore, but as a rule this metal occurs in particles too small to be seen with the naked eye.

The value of the ore varies widely. That worth from $25 to $40 a ton is commonly referred to as good ore and some being stoped at the time of visit had a gross value of $400 a ton. About 75 tons of ore is hoisted daily, practically all of which is milled in the 20-stamp mill at the mine. The cost of mining and milling is between $5 and $6 per ton, which certainly is not excessive when it is considered that the mine is about 25 miles from the railways and that the soft, heavy ground requires much timber, which is used in the form of squared "Oregon pine" (fir).

The Boston group of the Quartette Mining Company was extensively prospected some years ago, but the attempt to find workable ore bodies was unsuccessful. There are two shafts on the property, an incline about 300 feet deep and a well-constructed vertical shaft 400 feet deep. Both are now dismantled. The workings are close to the eastern border of the quartz monzonite, in much shattered and greatly altered andesitic porphyry, which breaks readily into small, dark-green or dark-red fragments and contains seams of epidote. The vein does not outcrop distinctly, but apparently strikes nearly east and west. Some small bunches of good ore were found, consisting of chalcocite, malachite, azurite, chrysocolla, and cerussite, but an average value of about $6 per ton was the best that could be obtained for any considerable part of the deposit.

DUPLEX MINE.

The Duplex mine is situated about half a mile north of the Quartette mine, almost on the southwest edge of town. It is on a strong fissure which, like that of the Quartette, strikes N. 65° W. and dips to the south. The material of the lode resembles in general character that of the Quartette vein. The mine was not producing at the time of visit, but had recently changed hands and is likely to be extensively developed in the near future. It was the first property in the district to ship ore, and the lode has been extensively stoped from the surface to a depth of 200 feet. Some of the shipments are said to have carried as much as $1,500 per ton. No attempt was made to examine the underground workings of this mine nor of the Good Hope mine, which is under the same management and lies between the Duplex and the Quartette.

SOUTHERN NEVADA MINE.

The Southern Nevada property includes two mines, one known as the Spokane, about 1½ miles north of Searchlight, and another known as the Blossom, about half a mile north of the Spokane. The 10-stamp mill and principal buildings of the company are all situated at the Spokane shaft, where the first development took place. Some good ore is said to have been found in this shaft, but when, shortly after this discovery, abundant water was encountered the shaft was utilized as a well and prospecting was begun on the Blossom claim. Here a large body of ore was uncovered, which was milled with water from the Spokane shaft, and produced about $325,000. The company is at present endeavoring to find a profitable continuation of this ore body.

The Blossom ore body forms a nearly horizontal blanket deposit that outcrops around the sides of a low rounded hill from 20 to 30 feet below its summit. This ore has been mostly stoped out so that the top of the hill is now supported merely by residual pillars. The rock composing the hill is a more or less altered porphyry, which seems in in some places to grade into the quartz monzonite and is probably merely a textural modification of the latter rock, which appears to be somewhat variable in the vicinity of the mine. Certain facies of this porphyry resemble so closely the "bird's-eye" porphyry of the Quartette mine as to suggest that something more than a hasty reconnaissance will be necessary to disclose the relations of the quartz monzonite, the intrusive porphyries, and the andesitic flows.

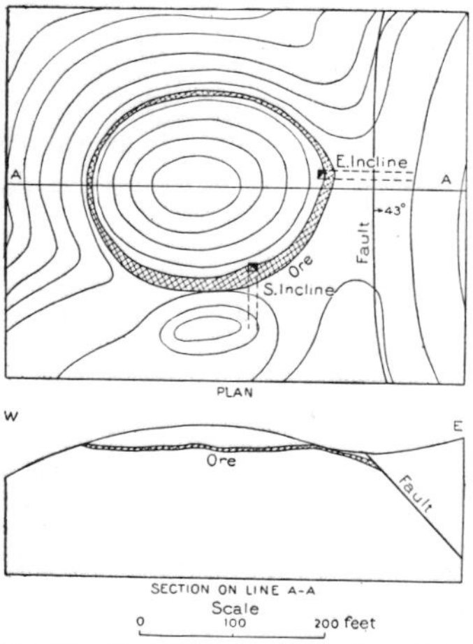

FIG. 15.—Sketch plan and section of the Southern Nevada blanket deposit.

The structural features of the Blossom ore body may be best understood by reference to the sketch plan and section of fig. 15. The main ore shoot measures about 300 feet from north to south and 250 feet from east to west. On the west and north it outcrops on the sides of the hill and beyond this outcrop has no known continuation in these directions. On the south and east, however, the ore comes to the surface and then dips under the slopes adjacent to the Blossom knoll. On the east an incline 430 feet in length has been sunk. For about 130 feet this follows what is clearly a continuation of the Blossom ore shoot, here showing a dip of 24° to the east. Below this point the incline steepens abruptly to 43° and thence follows a fissure of entirely different character. The steeper fissure carries no ore and shows much crushing and slickensiding, accompanied by a strong and persistent clay gouge. It is filled to a width of several feet with crushed and altered quartz monzonite, shattered quartz, and fragments of dragged-in vein material. It is apparently an unmineralized fault that has cut off the ore body in this direction. In spite of its unpromising character, drifts from 200 to 300 feet in length have been run on the 130, 230, and 330 foot levels.

The south incline (fig. 15) is 220 feet in length and follows the same zone of fissuring and mineralization as that in Blossom Hill. The dip at the incline is 35° to the south, but the drifts show that the lode curves in conformity to the contour of the hill, so that a southeast dip prevails east of the incline. The intersection of the lode and the north-south fault followed in the east incline has not yet been exposed.

Some $15 to $25 ore has been opened up in the drifts from the south incline, but this steep part of the lode is evidently of lower grade than that stoped from the horizontal part in Blossom Hill.

The Blossom blanket varies in thickness up to a maximum of about 3 feet. It is a zone of fissured and, in some places, shattered porphyry in which the rock fragments have been silicified, and the fissures and interstices, probably once partly filled with crushed porphyry, are now occupied by masses of porous, sugary quartz in which the individual grains are small, loosely cohering crystals. The quartz is in some places solid, but such material is usually barren. There are no definite walls to the deposit, and nearly horizontal branches are common. One of these has been followed in an incline to a point 50 to 60 feet below the main stope and at this depth shows a little good ore.

As a rule the ore is not visibly mineralized. Some of the best, however, shows free gold associated with limonite and little streaks of specular hematite. No sulphides were seen, the deposit being entirely oxidized. Some of the ore left unstoped is said to be worth as much as $55 a ton. The ratio of gold to silver varies, 1 ounce of gold to 4 of silver being apparently a common proportion.

The Blossom deposit is, so far as known, unique in the Searchlight district, although it is generally similar to some of the blanketlike lodes of Eldorado Canyon.

POMPEII MINE.

The Pompeii mine lies about 2½ miles nearly north of Searchlight and about half a mile north of the Blossom. The main lode strikes N. 65° W. and is supposed to dip about 45° SW. An inclined shaft 335 feet deep has been sunk in the foot wall of the lode at an angle of 58° and is being carried down to a depth of 600 feet. There is at present only one level, 265 feet below the collar.

The lode as exposed on this level is a brecciated zone in andesitic (?) porphyry, with a well-defined clay gouge on the foot wall and no definite hanging wall. The porphyry is much decomposed and oxidized, crumbling in many places like sand. Within the lode it is traversed by little veinlets of calcite and rusty quartz, many of them only a fraction of an inch in thickness. These little seams, particularly if soft and darkened by oxide of manganese, are said to carry such values as have been found in the lode. A crosscut, perhaps 25 feet in length, into the hanging wall shows a second lode containing considerable calcite, within which are embedded fragments of porphyry. Not enough work has yet been done to show the direction and extent of this lode, which apparently has no definite walls.

The Pompeii mine has produced no ore, although samples across 22 feet of the lode are said to have averaged $23 per ton.

Water is abundant below the 265-foot level and is increasing as the shaft goes down. At the time of visit a Cornish pump driven by a gasoline engine was raising 65,000 gallons in each twenty-four hours. Although this shaft was then 75 feet below the water level, the lode still maintained its oxidized character.

OTHER MINES.

Among the mines which could not be visited in the short time available are the Cyrus Noble, about half a mile west of Searchlight; the Searchlight-Parallel, about the same distance north-northwest of the Cyrus Noble; the Santa Fe, about two-thirds of a mile northwest of Searchlight, and the Searchlight mine, just south of the Blossom. The Cyrus Noble and Searchlight have recently erected 10-stamp mills and were preparing at the time of visit to hoist ore. The other properties are still in the prospect stage.

UNDERGROUND WATER.

When operations at Searchlight first began the only water near at hand was a small flow at Summit Spring, 2 or 3 miles east of the town. Later water was reached at 50 feet in Hall's well, near the spring, and at 60 feet in the Red Iron shaft, which was then utilized as a well. In the Spokane shaft of the Southern Nevada Company water was encountered at a depth of about 200 feet. The bottom of this shaft is only 4 feet below the original water level, but the supply has not noticeably diminished after about four years' pumping for milling and domestic purposes. In the Quartette mine water was reached 180 feet vertically below the collar of the shaft, and when the sixth level was opened over 12,000 gallons were pumped daily. The flow has increased with depth to the present time. The Good Hope shaft entered water at about 200 feet, and the Duplex at about 270 feet in vertical depth, the collar of the former shaft being 30 feet and of the latter 50 feet above that of the Quartette. The Pompeii shaft reached water 225 feet below the surface and at the time of visit was pumping 65,000 gallons daily. The water supply of the town is derived from a well 300 feet deep sunk in the quartz monzonite within the town. The bottom of the well is 60 feet below the water level.

The records show that in the vicinity of Searchlight, exclusive of the hills that rise notably above the town, abundant water may be expected at depths ranging from 200 to 300 feet below the surface, or in the neighborhood of 2,000 feet above the Colorado River. At points such as the Red Iron shaft, which is considerably below Searchlight in a canyon leading to the river, the water, of course, is reached nearer the surface. Deep mining will necessitate heavy pumping, but if the deep ores are not free milling they will nevertheless probably need concentration, and the water can thus be put to good use.

MINES OF ELDORADO CANYON.

DISTRIBUTION.

With the Eldorado mines are here included those of Nob Hill, situated on the main ridge 4 or 5 miles south of the canyon proper. Here active prospecting is in progress on the Silver Legion, Combination, and other claims, most of which were worked many years ago and produced good ore near the surface. None of the workings at the time of visit were over 100 feet in depth.

In Eldorado Canyon all the mines are on the north side. The principal group is situated near the head of the canyon and probably about 8 miles from the river. Here are the Quaker City, Mocking Bird, and Honest Miner. The first two properties were extensively worked many years ago, and the Honest Miner is said to have been the claim on which gold was first found in the district. The Quaker City and Mocking Bird are now idle, but the Black Hawk Mining Company is prospecting the Rand, Honest Miner, and other claims in the vicinity. All these mines are in fine-grained quartz monzonite.

About a mile lower down the canyon is the Wall Street mine, now idle, but with huge open stopes testifying to former activity. About 2 miles east of the Wall Street, or a mile east of Nelson, is the Buster mine, visible from the road down the canyon, and just over a ridge from the Buster, about half a mile northeast of it, is the Techatticup mine.

GENERAL CHARACTER OF THE DEPOSITS.

The ores at Nob Hill occur in east-west fissures in gneiss and schist. The dip is in most cases to the south and ranges from 70° to 90°. The veins are rather narrow and show no gouge and no evidence of movement since the deposition of the ore. In the Silver Legion mine the maximum width of the old stopes is about 5 feet. Although small, some of the veins are very regular and persistent, the Silver Legion, for example, being easily traceable over the surface for at least 5,000 feet. The sulphide ores, which in these veins are reached at 75 feet or less from the surface, consist of pyrite, galena, and sphalerite, in a gangue that is partly vein quartz and partly metasomatically altered gneiss or schist. The gold and silver present are probably inclosed in the sulphides, no native gold being visible. Most of the ore extracted in early days was oxidized, and contained free gold and horn silver. The Combination mine, 100 feet deep, is said to have produced about $150,000 from ore that was floated down the Colorado on flatboats.

The deposits near the head of Eldorado Canyon are notable for their diversity in dip. All are essentially mineralized fissure zones in quartz monzonite, but while some are almost horizontal and would ordinarily be called blankets, others are nearly vertical lodes. At the Quaker City there are two parallel, nearly horizontal veins about 100 feet apart, the upper one being exposed along the hillside for at least 1,200 feet. The average dip of these veins is 15°, in a direction a little east of north. They have been extensively stoped, in many places to a height of 7 feet, so that one can walk with ease over the smooth, slightly inclined floor of the great subterranean chamber thus formed. The ore consists of shattered quartz monzonite which has been cemented and partly replaced by calcite and quartz. Originally the deposit contained finely disseminated pyrite, but this is now oxidized and the material is a free-milling gold ore. In the Mocking Bird mine, about a quarter of a mile northeast of the Quaker City, the vein strikes N. 25° W. and dips northwest at an angle of 40°. It is in places 6 to 7 feet wide and resembles generally the Quaker City lodes. Quartz, however, is more abundant, and some parts of the vein are made up of solid white quartz inclosing fragments of quartz monzonite. This mine also contains large stopes, although development was apparently never pushed to any considerable depth.

Little could be ascertained concerning the value of the ore formerly extracted from these mines, but it is very improbable that ore worth less than $20 a ton could have been handled under the adverse conditions that must have prevailed when the mines in the canyon were most active.

On the Rand claim, near the Quaker City mine, a prospecting shaft is now being sunk and at the time of visit was 50 feet deep. The vein strikes N. 70° E. and dips north at an angle of 80°. The vein is tight, and consists of little bunches and veinlets of quartz, calcite, and pyrite in a somewhat pyritized quartz monzonite. The lode has produced some oxidized ore from old workings near the surface, but the value and dimensions of the sulphide ore bodies are yet to be determined. The Honest Miner claim, lying a short distance east of the Rand and Quaker City, is also being prospected. Here there are two nearly parallel northeast-southwest veins about 60 feet apart and dipping about 45° to the northwest. Some ore was formerly taken from the upper parts of these veins and treated in a 1-stamp mill with cyanide tanks improvised from barrels. The veins show the usual association of quartz and calcite with specks of limonite, and are in quartz monzonite.

The Wall Street mine lies a mile or two east of the Quaker City and Mocking Bird on an approximately east-west vein with a dip to the south ranging from 15° near the surface to 30° at a depth of 30 or 40 feet. A large body of ore extending from the surface down to the water level was stoped here in the early days of the camp. The distance from floor to back is in some places fully 20 feet. No timber was used, the back being supported by pillars of ore, and the open stope, resembling as it does a great natural cavern, is strikingly picturesque. The water level is apparently less than 100 feet below the surface. The ore, like that of the Quaker City and Mocking Bird, is shattered quartz monzonite, with bunches and stringers of quartz and calcite.

The Buster lode, formerly worked in a small way through shafts, is about a mile east of Nelson. The vein strikes northeast-southwest and has a steep northwest dip. The ore left on the dump shows quartz and calcite, with a little oxide of manganese and specks of malachite.

The Techatticup is one of the most extensively developed mines in the district. The vein strikes N. 80° E. and dips north at about 85°. It outcrops along the north slope of a steep ridge and is reached by an adit 180 feet in length and about 250 feet below the croppings of the lode. From this adit a drift several hundred feet in length has been run on the vein and at one point a winze has been sunk to a depth of 300 feet. Operations are at present confined to this winze. Above the adit level are three or four old adits and extensive stopes, which are in many places open to the surface. East of the main adit the lode divides into two, the southern branch running nearly east and west and dipping north at 60°. Both veins have been stoped along the surface to widths varying from 3 to 5 feet.

The vein is a regular and well-defined zone of shattered quartz monzonite in which the fragments are held in a network of quartz and calcite stringers. There has been no movement along the fissure since the ore was deposited, so that the vein is closely adherent to its walls and is almost impervious to descending solutions. Consequently, the sulphides at the surface, consisting of pyrite, sphalerite, and chalcopyrite, are only partially oxidized. The sulphides are usually rather finely disseminated. As the surface ore is said to have been rich in silver, galena is probably present in parts of the vein, although none was seen at the time of visit. The proportion of calcite varies widely and is greatest in that part of the lode lying east of the adit.

The average value of the ore formerly stoped from the Techatticup mine is said to have been about $40 per ton. Unoxidized ore from the bottom of the 300-foot winze is reported to have about the same value, but detailed or reliable information regarding this property could not be obtained on the ground.

In conclusion, it may be said that one of the most striking features about the Eldorado Canyon district is the fact that the Quaker City, Mocking Bird, Wall Street, and Techatticup mines, after having produced large quantities of ore under conditions less advantageous than those now prevailing, should be idle, or, in the case of the Techatticup, be prospected in a somewhat desultory manner. Of course the best ore may have been extracted, but the impression carried away from a hasty examination, without measurements or assays, was that the development of these mines under a single management and by modern methods constitutes a problem that is well worthy of careful consideration.

CRESCENT DISTRICT.

The little settlement of Crescent, consisting of a dozen or more tents and wooden buildings, sprang up early in 1905 in a district that was first exploited about ten years ago. The new town is 7 miles southeast of Nipton, on the road to Searchlight. It lies near the head of a small lateral valley opening on the northwest upon the desert plain traversed by the San Pedro, Los Angeles and Salt Lake Railroad and inclosed on other sides by barren hills a few hundred feet in height.

The general country rock is gneiss, in most places showing squeezed eyelike crystals (augen) of feldspar, associated with fine-grained schists and quartzite and cut by masses of granitic rock and by dikes of some basic eruptive now altered to amphibolite.

The principal active prospect is the Big Tiger, half a mile east of town. The so-called lode seems to be in reality a roughly semicircular mass of shattered quartzite that has been recemented by quartz and carries more or less gold and silver. The quartzite is apparently a residual patch, the remnant of a syncline, which rests upon the gneiss and is cut off on the southwest by a fault. Similar patches are reported northwest and southeast of the Big Tiger and presumably are similarly related to the same fault. The quartzite is much shattered by repeated movements, shows in places a schistose structure, and may be of the same general age as the gneiss.

The best assays have been obtained near the fault and indicate values of $10 to $12 per ton. The deposit has been prospected upon the assumption that it is a wide, irregular lode, and no attempt has been made to explore the fault, whose presence seems to be demanded by the structure, although the actual fissure is not clearly exposed. A shaft 100 feet deep has been sunk in the quartzite a few feet east of the fault, and its bottom is apparently very near the base of the quartzite. The fissured and crushed quartzite is all oxidized and shows no indication of ore other than a little iron oxide and a few specks of malachite.

About 3 miles southwest of Crescent, on the other side of a gneiss ridge, is the Calavada property. Here also the deposit appears to be a much shattered and silicified bed or beds of quartzite resting on the gneiss and dipping gently south at an angle very little steeper than the general slope of the ridge upon which the deposit lies. The quartzite mass is fully 1,000 feet wide from east to west, and the distance from the highest point on the slope to the point where the deposit passes beneath the alluvium of the desert is probably still greater. A shaft 40 feet deep has been sunk at one place without reaching the underlying gneiss.

The material is much shattered and in many places is a breccia of quartz fragments. It is generally rusty and varies in hardness. The deposit has been carefully and conservatively sampled over an area about 1,000 feet square for the purpose of determining its availability for extensive cyanide treatment, the proposition being to handle the material with steam shovels. These assays rarely fall below $2 per ton, and some of them run as high as $12. The average for the whole area is approximately $3 per ton. The deposit is separated from the railway by 2 miles of gently sloping plain, and water can be pumped from wells sunk in a playa or "dry lake" less than 3 miles away.

The occurrence of the gold in the quartzite of the Crescent district is probably due to the favorable conditions for deposition afforded by the shattering of beds too brittle to conform without fracture to movements that merely folded or crumpled the underlying gneiss and schist. Part of the brecciation may be due to faulting. In the case of the Big Tiger, the gold-bearing solutions may have come up through the fault previously mentioned. The source of the gold in the Calavada property, however, is not apparent, although the great disturbance of the mass would indicate that here, too, fault fissures may have provided channels for the rise of auriferous solutions.

GOLD MOUNTAIN DISTRICT.

SITUATION.

The mines and prospects of the Gold Mountain district, which are being worked at present, are situated on the eastern slope of the Grapevine Range, 20 miles northeast of the head of Death Valley and 11 miles west of Montana, a stage station halfway between Goldfield and the Bullfrog district. The old town of Gold Mountain, on the Death Valley versant of the range, and Tokop, about 5 miles to the north, were not visited in the single day devoted to this region.

GEOLOGY.

The higher points of the rather irregular hills that here represent the Grapevine Range rise steeply from 1,500 to 2,000 feet above the alluvial slopes at their eastern base, and are composed of an intrusive mass of biotite granite, which has a width of 2 or 3 miles from north to south and extends for an unknown distance westward. This granite is intrusive into a series of sediments, originally shales with lenses of limestone, and probably of Cambrian age. The shales are metamorphosed into fine-grained, rather obscurely crystalline biotite schists with subordinate muscovite and hornblende schists, and the limestone is recrystallized. The metamorphism is probably due to the intrusion of the granite and becomes less intense with distance from the eruptive rock. Both granite and schists are cut by numerous dikes of aplite and by dikes of fine-grained hornblendic rock, probably diorite porphyry. The metamorphic rocks form low hills to the north of the granite peaks and are overlain by a thick series of Tertiary volcanic rocks. At the base of this younger group are flows of rhyolite with intercalated tuff beds, the whole being over 500 feet thick. Overlying the rhyolitic flows and tuffs, which seem to have filled a basin and thin out to the west, is a bed of red tuffaceous material, containing abundant fragments of granite. This bed is conformably overlain by a heavy capping of olivine basalt. In some places the Tertiary volcanic rocks are evidently dislocated by normal faults of considerable throw.

ORE DEPOSITS.

The ores, which are gold ores with subordinate silver, are found in veins traversing the granite and schists. The deposits appear in most cases to have been originally solid tabular masses of quartz; but all show more or less brecciation due to movement since the fissures were filled. None of the workings has yet gone below the zone of oxidation, although pyrite, chalcopyrite, galena, and tetrahedrite are frequently found in the generally oxidized material.

The principal mine in the district, the Rattlesnake, owned by the Bonnie Clare Bullfrog Mining and Milling Company, is in schist about 2 miles northeast of the main granite ridge. The mine has been worked intermittently for over ten years, part of the ore having been treated in a 5-stamp mill at Thorps Well, about 2 miles south of Montana station. In all, from 3,500 to 5,000 tons of ore have been run through the mill. The mine is opened by two adits about 125 feet apart vertically. The lower adit is over 400 feet long and connects with a winze 200 feet deep, from the bottom of which extend drifts about 100 feet in total length. Most of the ore formerly mined came from the upper tunnel, the stopes extending in places to the surface. At the time of visit the mine was temporarily idle.

The Rattlesnake lode strikes nearly east and west and dips north at 75°. The country rock consists of thin-banded slaty schists, some of the bands evidently representing altered limestone or calcareous shale. The planes of schistosity, which in general appear to correspond with the original bedding, dip south at various angles, the average inclination being about 45°. The lode thus cuts the planes of schistosity at a considerable angle.

The vein has a maximum width of 4 feet. It is a fairly regular filled fissure and in places shows banding due to crustification in the quartz. Like all the veins seen in the district, however, it has been affected by later movements and is generally brecciated. It consists of quartz, somewhat stained with oxide of iron. A streak of fine soft specularite, resembling rouge, occurs along the foot wall in the upper tunnel. No sulphides were seen at the time of visit, although a little galena is said to be occasionally found. The calcareous bands in the schist have apparently exerted no influence upon the vein itself, which is free from calcite. This mineral, however, is abundant in stringers and vugs in the walls in or near the limestone bands. The value of the ore, as shown in an apparently truthful statement published by the company and based on 68 samples taken from various parts of the workings, ranges up to $213 a ton, the average being approximately $23. No account was taken of the silver in these assays, the amount of this metal present rarely exceeding 2 ounces per ton and being usually only a fraction of an ounce.

About 1½ miles southwest of the Rattlesnake are the properties of the Nevada-Goldfield and Nevada-Sunshine mining companies, operated under one management, and locally known as "Parson's camp." Here active prospecting is in progress and a vertical shaft is being sunk to a depth of 300 feet in order to explore at that depth veins upon which considerable work has been done by tunneling. There are two principal veins, both in granite and about 450 feet apart. They strike about N. 40° W. and dip northeast at angles varying from 70° to 75.° Most of the work has been done on the southwest vein, which in places is fully 20 feet wide, and is said to have been traced on the surface for a distance of nearly 2 miles. This vein, which outcrops along the steep north side of the main granite ridge, has been opened by three tunnels, with several hundred feet of drifts, and by a shaft and winze at the lowest point of outcrop to a depth of 180 feet. The new shaft is being put down midway between the two veins.

The southwest vein consists of shattered quartz mingled with crushed granite. Two persistent seams of gouge are usually referred to as the hanging wall and foot wall, although the original fissuring and veining of the granite extends in many places beyond these so-called walls, which are due to later movement. In a few places the vein has been offset from 12 to 15 feet by normal faults.

The best ore occurs usually near one of the walls, in quartz less shattered than the mass of the vein, which, as a whole, shows no conspicuous mineralization. The ore commonly shows stains of malachite and contains some cerussite and pyromorphite. In places a little galena and a gray copper mineral, probably tetrahedrite, are found associated with the prevailingly oxidized ore. Free gold is rarely seen and no native silver has been noted. Selected samples of some of the ore have afforded assays up to $4,600 a ton, but the vein as a whole, to the depth thus far explored, is certainly of low grade. No ore has yet been shipped.

About half a mile northwest of Parson's camp is the Oriental mine, no longer worked. This was probably the first mine opened in the district and was operated through some small inclined shafts. A single loose mass of ore from the surface is reported to have been shipped to Carson City many years ago and to have produced over 50 ounces of gold. The vein, which is in schist, strikes east and west and dips north. It is well defined at the surface, and is of the brecciated type characteristic of the district.

The Courbet mine, under the same ownership as the Rattlesnake, is situated about a mile west of the Nevada-Goldfield property. It has produced some ore and over 300 tons, said to average $32.50 a ton, is piled on the dump. The mine was idle at the time of visit.

The development consists of two adits about 75 feet apart, the upper one being 350 feet and the lower one 400 feet in length. Both are on the general line of the vein, which strikes nearly north and south with an easterly dip of 30°. The vein, which is in granite, was originally a fairly regular plate of quartz up to 15 inches in width, carrying pyrite, chalcopyrite, galena, and perhaps other sulphides. Movement since these minerals were deposited has crushed the quartz and produced seams of gouge along the hanging wall. The crushed vein material is all more or less oxidized and contains abundant cerussite, with malachite, azurite, and pyromorphite. The ore is mainly a gold ore, although it must contain notable proportions of silver and lead.

The Gila Monster mine lies about 2 miles south of the Nevada-Goldfield, on the south slope of the same granite ridge. It is owned by the Butte Gold Mining and Reduction Company. The lode was discovered in February, 1905, and is being actively prospected by a tunnel on the vein, about 400 feet long at the time of visit, a shaft 100 feet deep above the tunnel, and a winze about 50 feet deep near the portal of the tunnel. The vein strikes east and west and dips north at 45°.

The vein is large, the whole fissured and mineralized zone being from 15 to 20 feet wide, and it can be followed along the steep hillside for several hundred feet. The original fissuring appears to have followed a dioritic dike in the granite and the lode probably consisted of one or more thick plates of quartz, with many subordinate stringers and no very definite walls. Later movement has shattered and crushed the original vein material. The ore is generally oxidized, but contains a little pyrite and galena. Copper minerals are not as abundant as in the Nevada-Goldfield and Courbet veins.

NOTES ON THE MANHATTAN DISTRICT.

By W. H. Emmons and G. H. Garrey.

INTRODUCTION.

Manhattan, Nye County, Nev., is in latitude 38° 28′ N., about 29 miles north-northeast of Tonopah and about 13 miles southwest of the old silver camp, Belmont.[a] It is on the western slope of the low group of hills locally known as the Smoky Mountains, which connect the Toquima Range on the north with the San Antonio Mountains on the south and separate Big Smoky Valley on the west from the northern arm of Ralston Valley on the east. The nearest railroad point is Tonopah, with which it is connected by daily stages and automobiles.

In March, 1906, a hasty reconnaissance of the geology and mines of the district, incidental to other work in Nevada, was made jointly by the writers. At that time a large part of the country was covered with snow, and much of the surface geology was masked; however, in view of the prominence which Manhattan and the surrounding country has recently attained, it seems desirable to record such facts as were observed.

TOPOGRAPHY.

The town of Manhattan has an altitude of about 7,250 feet above sea level, and is situated in a narrow valley followed by the old Belmont-Cloverdale wagon road, which has been one of the principal highways of the desert ever since the early sixties. The rounded hills on either side of this valley rise only 200 to 500 feet above its floor, but about 1½ miles to the northwest they grade into the rugged mountains forming the crest of the Toquima Range, which extends northward about 80 miles farther, paralleling the Toyabe Range to the west and separated from it by the Big Smoky Valley.

FUEL AND WATER SUPPLY.

Owing to its greater elevation, Manhattan possesses advantages not enjoyed by mining camps in the desert country farther south. The hills surrounding the town are covered with vegetation and support a growth of scrub pine and cedar that is of great value for fuel and mine

[a] Location furnished by Mr. Arthur Philbrick, United States deputy mineral surveyor, Manhattan, Nev.

timbers. Good water, sufficient for present purposes, is obtained from relatively shallow wells in the valleys, and it is believed that this supply can easily be augmented by running pipe lines from springs in the mountainous country to the southeast.

HISTORY.

Gold was discovered at Manhattan by John C. Humphrey and companions in April, 1905, near the southern base of April Fool Hill, about 100 feet from the Belmont-Cloverdale wagon road. The first ore found gave only low assays, but in July specimens of high-grade ore showing an abundance of free gold were obtained. Interest increased as subsequent discoveries were made, and in August there was an influx of prospectors, who located most of the ground for several miles around. Attention, however, was soon drawn to other fields, where new strikes were reported, and consequently early in December the town had less than 100 inhabitants. A shipment of rich ore by way of Tonopah caused an excitement there, and people from there and from near-by camps, waiting for such an opportunity, were quick to rush to the new camp. Within a few weeks a town a mile in length was built in the valley and its tributaries, and in March, 1906, there were 3,000 people in Manhattan and the immediate vicinity. The town is agreeably located, well laid out, and has many wooden buildings, while some of stone are in process of construction. It is provided with post-office, numerous stores, banks, newspapers, assay offices, telegraph and telephone service, and stage and automobile lines.

DEVELOPMENT AND PRODUCTION.

At the time of the visit, in March, 1906, most of the claims were being vigorously prospected and several mines were shipping ore. Much of the ground had been let out to leasers for a period of six to twelve months, each lease having 100 feet along the lode and extending to the side lines of the claim. The leasers usually pay a royalty of 25 per cent of the gross receipts. Briggs & Evans, who owned lease No. 14 on Union No. 9 claim, had sunk a shaft on the vein to a depth of 163 feet. Most of the openings, however, were less than 50 feet deep. From many of these shafts drifts extended for a distance of 10 to 50 feet.

The production, so far as could be learned, was about $50,000. On account of excessive freight rates and treatment charges ore averaging less than $70 was left on the dumps, awaiting the completion of mills.

GENERAL GEOLOGY.

The rocks around Manhattan include slates, limestones, quartzite, granite, diorite porphyry, and rhyolite. The ore bodies so far discovered are in the sedimentary rocks.

SEDIMENTARY ROCKS.

The slates, limestones, and quartzites which make up the sedimentary series at Manhattan are exposed for 2 or 3 miles along the Tonopah road and perhaps as far on the road to Belmont. Mr. S. F. Emmons[a] describes a series of slates, limestones, and quartzites at Belmont closely resembling these. Further, Mr. G. K. Gilbert[b] notes that the slates at Belmont contain Silurian fossils, and estimates their thickness at from 4,500 to 5,000 feet. The slates at Manhattan are probably of the same age.

Near Manhattan, as at Belmont, the sedimentary rocks are extensively metamorphosed and consist for the most part of dark, fine-grained mica schists or phyllites. The limestone layers are in most cases less than 30 feet thick, and probably will be found not to persist for very great distances. Usually they have been entirely recrystallized and in some places changed to an impure gray marble. Quartzite layers also occur interbedded with the belts of schist. A quartzitic breccia or conglomerate layer less than 20 feet thick, made up chiefly of large subangular quartz fragments, occurs near the Reilly fraction, dipping with the schists. At Manhattan the glossy or satin-like phyllites or fine mica schists predominate over other members of the series. These are present in considerable variety, but the commonest facies is one which in hand specimens appears to be composed almost entirely of biotite. Microscopic study shows that this schist consists chiefly of biotite, muscovite, and quartz. The mica forms interlocking folia in perfect alignment with the schistosity, inclosing small grains of quartz having also a pronounced though less perfect linear arrangement. In certain facies crystals of chlorite and magnetite cut across laminæ of the mica in a manner which shows that they were formed after the schistosity was produced. The schistosity as a rule is approximately parallel to the bedding, as determined by contacts of limestone and slates.

GRANITE.

Granite outcrops about 3 miles southeast of Manhattan along the road to Tonopah. It is coarsely crystalline, pink or gray, and consists chiefly of quartz, feldspar, and mica. A similar body of granite occurs to the northeast of Manhattan on the road to Belmont. Mr. S. F. Emmons,[c] describing relations of granite and sedimentary rocks at Belmont, says: "The metamorphic rocks where in contact with the granite are frequently highly altered and may come under the general name of quartzite." Mr. Spurr[d] also states that the metamorphism of the sedimentary rocks to the southwest of Belmont is connected

[a] U. S. Geol. Explor. 40th Par., vol. 3, p. 394.
[b] U. S. Geog. Surv. W. 100th Mer., vol. 3, p. 180.
[c] Loc. cit., p. 396.
[d] Bull. U. S. Geol. Survey No. 208, p. 91.

with the intrusive mass of granite. It is highly probable, then, that the granite in the vicinity of Manhattan is also intrusive in the sedimentary rocks and may be regarded as one of the principal causes of their metamorphosed condition.

DIORITE PORPHYRY.

About 1½ miles northeast of Manhattan there is a large body of diorite porphry in contact with rhyolite. The contact occurs along the bottom of a small gulch for a considerable distance, but when this place was visited 2 or 3 feet of snow lay in the bottom of this gulch, and the relations of the porphyry and rhyolite could not be clearly made out. The porphyry is undoubtedly of later age than the slates, but whether it is contemporaneous with or of later age than either the intrusive granites or the rhyolites has not yet been determined. The diorite porphyry is brown on the weathered surface and shows a large number of light-colored feldspar phenocrysts. On a freshly fractured face the groundmass is dark green and has phenocrysts of about the same shade. Microscopic study of an altered specimen shows that the porphyry consists essentially of plagioclase feldspar (andesine and labradorite) with a large amount of calcite, serpentine, and chlorite. The serpentine and chlorite are probably the alteration products of hornblende or pyroxene, and the rock is an altered diorite porphyry.

RHYOLITE.

Rhyolite occurs as surface flows covering both the sedimentary rocks and the granite and also as dikes cutting the same formations. To the north of Manhattan these flows cover a vast area extending approximately to the summit of the Toquima Range, giving a vertical thickness of many hundred feet. North of Belmont this range, according to Mr. Spurr,[a] appears to be entirely covered by these great flows of rhyolite, which are also found in the San Antonio Mountains to the southward. The rhyolite is commonly yellow, pink, or light gray and contains many phenocrysts of smoky quartz and white feldspar, with occasionally a small flake of mica. The basal facies very often contain a large number of fragments of schist, granite, and rhyolite. Under the microscope the groundmass exceeds the phenocrysts in volume and is usually glassy, sometimes spherulitic, and frequently devitrified. The feldspar crystals are orthoclase and acid plagioclase. The quartz phenocrysts usually show embayments, due to magmatic corrosion.

The rhyolites are eruptive rocks, presumably Tertiary, which are so extensive that they completely mask the earlier rocks over the larger portion of the Toquima Range.

---

[a] Loc. cit. p. 92.

FAULTING.

That there has been extensive faulting in the ore-bearing rocks is shown by brecciated zones and polished striated walls exposed by mining explorations. Fractures traverse the slate, limestone, and quartzite layers in all directions, and along some of them the beds have been displaced. In one or two cases fissure veins have been slightly displaced, which shows that some faulting also occurred after the mineralization.

About three-fourths of a mile north-northeast of Manhattan a fault striking N. 50° E. and dipping southeast at 55° brings the rhyolite flow breccia into contact with the truncated edges of sedimentary beds. Movement striæ on the slickensided surface dip northeast at 75°. At a point farther northeast on the strike of this fault rhyolite forms both the foot and the hanging wall, which shows that this fault was formed subsequent to the extrusion of the rhyolite.

ORE DEPOSITS.

GENERAL CHARACTER.

So far as developments have shown the ore deposits of the Manhattan district are confined entirely to the area of metamorphosed sedimentary rocks, which is surrounded by rhyolite, granite, and alluvium.

The ore bodies are (1) narrow, tabular lode deposits, occurring in rather continuous veins crosscutting the bedding and schistosity of the metamorphosed sediments, (2) deposits interlaminated with the schistosity, and (3) deposits parallel to bedding planes of the quartzites, shales, and limestones and replacing part or all of limestone or calcareous layers.

FISSURES.

Most of the principal ore-bearing fissures of the district, so far as could be learned from a considerable number of compass readings, strike north or a few degrees west of north. Some mineralized fractures, however, also make considerable angles with this main belt of veins. There are also numerous minor fractures, only slightly mineralized, which form a network of intersecting planes running apparently in nearly every conceivable direction.

The dips of the principal veins range from 45° to 90°; but a majority of them are inclined to the southwest at angles ranging from 60° to 70°. On account of their undulating character some of the veins may also vary considerably in strike and dip at different points along their course.

THE ORES.

The chief gangue minerals of the ores are quartz and calcite, but barite and fluorite are sometimes present. Gold is the principal and almost the only metallic constituent of value, although silver is present at some places in amounts varying from a trace to $2 a ton. Small quantities of ore carrying argentiferous galena have been reported as occurring near Broncho Hill and also in the neighborhood of Central City. Iron and manganese oxides are usually present, but are of no economic value. The gold occurs usually as minute particles, too small to be detected in the hand specimen, finely disseminated through the quartz, calcite, or claylike gouge filling the veins, though there is occasionally coarse gold in small transverse stringers and small sheets of leaf gold in cracks along bedding planes or between the laminæ of the crushed schists. Coarse gold, however, is an exception and usually occurs only in small bunches or pockets, very irregularly and erratically distributed. On the Mayflower and Annie Laurie claims specimens of gold-bearing quartz were found which had a flaky or schistose structure, as though a calcareous schist had been replaced by silica particle by particle.

The typical rich ore of the camp is either a gold-bearing dense white quartz, stained in places to a yellowish color by iron oxides, or a mass of porous white quartz formed by intergrown tabular crystals of quartz which are pseudomorphs after calcite. In certain instances this ore is associated with a dark-brown powder which is a mixture of iron and manganese oxides. The gold in such ore is usually not visible, even with a magnifying glass, but on crushing finely and panning a good "streak" of very fine bronze-colored gold is obtained.

The shipments of this class of ore so far have yielded returns varying from $70 to $300 or more to the ton, but the average value has been about $125. Numerous assays of picked specimens gave values running from several hundred dollars to many thousands of dollars to the ton.

A medium grade of ore is also found where small breccia fragments of the country rock have been cemented into a compact mass by dense white quartz similar to the quartz forming the best ores. This class of ore usually ranges between $35 and $70 and hardly justifies shipment at the present high rates for haulage and treatment.

The ore too low to warrant shipping is classed as milling ore. The type of low-grade ore common to the fissures which cut across the formations consists usually of a breccia of schist fragments loosely cemented by a porous matrix of quartz or of a kaolinized or clayey substance resulting from the alteration of the crushed wall rocks. At a few places the country rock immediately adjacent to the fissures, it was reported, carries values sufficient to warrant its being mined as milling ore. The values in this case are probably not usually due to a fine dissemination of the ore-bearing minerals through the rocks, but to thin films of ore along the numerous cracks intersecting the fractured country rock.

High-grade ore may also be obtained by sorting out portions of the veins in which the limestones or calcareous shales have been entirely or partially replaced by gold-bearing quartz. Other portions of the limestone layers which have been replaced to a less extent by quartz are leaner and are classed as "mill dirt." The milling ore on the dumps is reported to run from $10 to $35.

In getting ore ready for shipment most of the material from the vein is run through a screen and all the "fines" are shipped directly, while the balance is sorted and the fragments of comparatively fresh wall rock, free from quartz, are picked out, for these usually carry no values.

Treatment charges (including reduction for moisture) and haulage together amount to $35. As a rule, however, no ore which assayed less than $70 a ton was being shipped.

LODE DEPOSITS.

The lode deposits consist of roughly tabular mineral masses along well-defined fissures which cut across the schistosity and the bedding planes of the metamorphosed rocks.

Typical examples of this type are the ore bodies of the Stray Dog, Union No. 9, Little Grey, and other claims in the same vicinity. In deposits of this kind the principal part of the vein consists chiefly of a breccia of schist, quartzite, and limestone embedded in a matrix of pulverized and altered country rock which in places has been partially or entirely replaced by silica and calcite.

The richest portion of the deposit is usually a dense, hard, white quartz, frequently stained to a yellowish color by iron, or a belt of "honeycombed" porous quartz resulting from a pseudomorphic replacement of vein calcite and barite by silica. These zones of rich quartzose ore, which vary in width from 1 or 2 inches to 2 or more feet, grade into belts of lean ore that contain abundant breccia fragments of schist and other wall rocks and have a clayey material cementing the breccia fragments in place of the quartz. This lean vein material, which is in places several feet wide, then grades into crushed country rock which often carries small values to a considerable distance from the main fracture. At present the development is not sufficient to throw much light either on the shape of these ore bodies or on their distribution along the vein. Nevertheless, judging from somewhat similar deposits in other areas, it is probable that the pay shoots of irregular size and shape will be separated from one another by portions of the vein which are lean or barren.

INTERLAMINATED DEPOSITS.

Such deposits occur immediately southeast and also north of town, and consist either of narrow veinlike deposits, resulting from mineralization along planes of movement parallel to the schistosity, or of isolated and comparatively small lenticular bodies of mineralized quartz which have their long dimensions oriented with the laminæ of the schist.

The narrow veinlike deposits are variable in extent and erratic in distribution, owing to the tendency of the veins to pinch to a mere crack and then to widen out again. The exploration for the pay shoots is made difficult not only as a result of the pinching of the veins, but also because the ore streak at some places plays out along one vein and continues across a series of smaller transverse fractures to a parallel vein several feet away and then proceeds along the latter in the same direction. These deposits consist of a porous or honeycomb quartz associated with the gouge resulting from the crushing and trituration of rocks due to movement along the planes of weakness parallel to the schistosity. The ore probably partially results from a substitution of auriferous quartz for the recrystallized calcitic constituent of gouge formed from very calcareous schists or by a quartz replacement of the calcareous schists immediately adjacent to the fracture.

The small discontinuous lenticular quartz bodies parallel to the schistosity are often overlapping, although separated by considerable areas of schist, the laminæ of which bend around the lenslike masses and inclose them on all sides.

At some places splendid specimens of leaf gold occur between the laminæ of the schist. Considering, however, the amount of unprofitable exploratory work necessitated by the small size and irregular distribution of the interlaminated deposits, it is more than likely that this form of deposits will prove less remunerative than those in the crosscutting fissures.

BEDDED DEPOSITS.

The bedded deposits are roughly tabular and conform with the bedding of the country rock. Their boundaries are seldom regular in shape, and they vary in thickness from a few inches to 20 feet or more. Most of them have resulted from the partial or complete replacement of beds of finely crystalline limestone by silica, while some are belts of mineralized, crushed, and pulverized country rock, brownish-gray gouge, and quartz, following the bedding and resulting from extensive movement. Some of these belts are of considerable thickness and give satisfactory values.

During the process of replacement of the limestone or calcareous shale the waters carrying the silica in solution perhaps also brought in the precious metals, which seem to have been deposited along with the silica, since the quartz appears to carry most of the values. Apparently before the limestone was replaced it was completely recrystallized to a cleavable mass or aggregate of large white to colorless calcite crystals, the incomplete outlines of which at many places grade irregularly into the massive limestone. Subsequently silica-bearing waters replaced the calcite crystals. Specimens collected in the Manhattan Consolidated mine illustrate well the various transition stages in the replacement of individual calcite crystals by quartz. In some specimens there appears to be only a thin film of quartz enveloping a part or the whole of a crystal. Again, the silica is found to have penetrated irregularly for short distances along a few of the cleavage planes of the calcite, while in other specimens nearly all the numerous cleavage partings are filled with silica, although calcite remains between the thin siliceous films. Still other specimens show that the calcite which formerly intervened between these films has been partially or entirely replaced by silica. In many instances the siliceous pseudomorphs after the calcite are rather porous, the silica having failed to replace all of the calcite taken into solution.

The replacement of calcite by silica has proceeded most extensively along the contacts of the limestone with other formations, or along bedding planes and joints in the limestone itself.

DISPLACEMENTS ALONG THE FISSURES.

A large number of fractures along which movement has occurred traverse the metamorphosed sedimentary rocks.

Evidence of movement is found in the soft, claylike "gouge" which at some places fills both barren and productive fractures. The gouge in one or two instances reached a width of nearly 4 feet, while the country rock was crushed and brecciated for several feet on either side of the fracture. Polished or slickensided surfaces showing steeply inclined (50° to 75°) movement striæ are also of frequent occurrence.

On account of the snow and the limited time devoted to the study of the area it was impossible to determine the amount of displacement resulting from the movements, but by detailed mapping of the most prominent and easily recognizable limestone, quartzite, and shale beds both the direction of the faults and the amount of displacement of the beds can probably be ascertained.

To what extent faulting has occurred since the deposition of the ore is at present undeterminable on account of the limited exploration along even the more promising veins. However, a fault which has caused a displacement of only a few feet cuts one of the mineralized fissures on the Stray Dog claim. Again, near the southeast boundary of Union No. 9 claim there is an irregularity in the pay streak in which the north portion of the vein appears to have been offset 6 feet to the east. The present workings do not show conclusively, however, whether this is a fault in the vein or two parallel veins 6 feet apart with a small transverse fracture connecting the ends of the pay streaks on the two different veins.

The faults are probably normal in character, which implies a relative downward movement of the hanging wall. However, there was undoubtedly also considerable horizontal movement or thrust along the fault plane, as shown by striæ and grooves. There appear to have been two or more periods during which faults were formed. The chief fracturing was probably connected with the intrusion of the granite, though there was faulting after the eruption of the rhyolite and also possibly after the intrusion of the diorite porphyry.

"Struck it at Last"

The special interest of this illustration lies in the fact that it is reproduced from a genuine photograph of a genuine prospector

THE NATION BACK OF US, THE WORLD IN FRONT

OUT WEST

Vol. XXVI No. 2 FEBUARY, 1907

THE NEVADA BONANZAS OF TODAY

By WILLIS GEORGE EMERSON.

STRIPPED of every contour that would please the eye, homely and unattractive, Nevada lies today the flat chess-board upon which is being played a game of elemental passions that is holding the attention of the entire civilized world.

The desert state as the beacon light, has first attracted and then dazzled 50,000 men. The figures in the foreground of this remarkable picture of human interest and rampant greed are as unique and striking as in those traditional days when the miner was a millionaire or pauper, according to his ideas, sometimes within an hour. Three years ago the sage-brush state was living upon a memory. Its little capital of Carson was filled with old-time citizens who gained a livelihood from the one productive spot in the entire state and told over and over again the never-to-be-forgotten story of the old Comstock. As the levels in this enormous historic mine, which produced eight hundred millions of dollars and gave with a free hand the money to save the Union, sunk deeper and deeper into the bowels of the earth, the richness of the ores gradually became less and less and the terrific heat, which is 170 degrees on the lowest levels, was rendering the work more and more difficult.

It seemed as if, with the gradual dwindling of this gigantic mine, the old state of Nevada was shrinking and dwindling, too. Occasional prospectors left the capital of the state and were gone for a week or a month, only to return with bitter stories of the hardships they had endured and with disconsolate tales of the barrenness of the rocks.

When all seemed at a standstill, a straying mule from the camp of a prospector named Jim Butler, scrambling up a hill, broke a piece of rock from a ledge, but a short distance from the camp-fire. The aspect of the broken face of the rock attracted the attention of the miner and he threw it into a sack to be assayed upon his return to civilization. The assay gave astounding returns, and, when its results became known, the town of Tonopah sprung to life.

Today the state of Nevada teems with energy. It is busy, it is rich, it is rough and ready, it is vital, and it is most keenly alive. Towns have sprung up like magic over the heretofore uninhabited plains. Dug-outs, tents and adobe houses were the first progenitors of this new civilization. Now the streets of the older mining camps are backed by substantial testimonials of the progress of recent years. But the spirit of '49 is there still, as evidenced in the saloons, whose rough signs, changing here and there to the planed and painted boards of maturer growth, still jet out over planked sidewalks, calling attention to the liquors that may be obtained within. Small buildings lie in juxtaposition to stone edifices reared as testimonials to the sudden fortunes that have been acquired by their owners. Gambling rooms, general provision stores, brokers' offices, laundries, bank buildings, and livery stables are all mixed and shuffled together like cards in a gambler's hands. Many of the saloons look as if they might be left over from the days when a man stood up with his back against the wall and the bowie knives, from artists in their use, whizzed by within half an inch of his head, the winner of the game being the one who could throw the closest without really hitting.

Throughout the days and nights the roar of the blast and the rattle of machinery is heard from the mines. On the streets, great freight wagons trek through the dust, hauled by teams of forty mules, setting out on their destination from the industrial centers of the state toward the younger camps springing up in the distant horizon. When the night falls, the different "joints" flare with light, and the click of the ivory balls in the roulette wheels, the rattle of dice and the flip-flap of cards is borne to the ears of those strolling by on the street. Throughout the state, the life of the mining camps is much the same. With Tonopah possessed perhaps of the more advanced civilization and Goldfield crammed with the most intense energy, Carson City, the capital of the state, is the oasis of the entire desert. Its streets are dead and quiet after nine o'clock. It is the ideal home-site where those in the foreground of the battle of wealth house their wives and children. In the strife of its pet railroad, the old Virginia-Truckee, which carried the famous miners of the past years, has been cut off from communication by the Southern Pacific, which in contempt of its rival heads directly south from Hazen, relegating Carson to the position of an out-of-the-way stopping place.

The mining camps of the desert state are the new blood that is filling its veins. They sweep in a wide circle, radiating like the spokes of a wheel, with Goldfield as the hub. Their name is becoming legion and their wealth is fabulous. Fairview, Tonopah, Manhattan, Wonder, Yarington, Bullfrog, Rhyolite, Searchlight, Beatty, Lida, Bonnieclare, and last but not least, Greenwater.

Goldfield is the center of them all. It is the driver's seat from which the reins run out that guide the vehicle of state. It is an anomaly; it is old and it is new. The government is ridiculously simple—the Sheriff, the District Attorney and the Justice of the Peace run the town. On the other hand, it has a completely organized fire department which throws grappling irons over wooden houses and jerks them into a place of safety in case of fire. Also,

Photo by Dickerson

AN AVERAGE BIT OF SOUTHERN NEVADA

somewhere in the archives of six months past, rest the forgotten papers of a Chamber of Commerce.

In contra-distinction to every other mining camp in the world, automobiles are the most common means of travel for the prospector, and the sight of a miner speeding over the hills by means of electric power is but a touch of the individuality which marks this town for its own. Fully 12,000 souls have flocked to Goldfield seeking to find the realization of life-time hopes. Others are crowding in. The population is made up of a composite of the riff-raff of the world and of the most civilized. The shrewdest brains of the criminal fraternity have foregathered like vultures seeking their prey. Investors, capitalists, curiosity hunters, gentlemen of leisure, all throng the streets and jostle one another on the narrow sidewalks. Some of these men have discovered that for which they have been seeking and are more or less satisfied. Like Antæus, they have touched the earth and been strengthened; or, emulating the stock frenzy of Wall Street, they have dabbled in mining shares, whose true basis of worth is in this case founded upon the gold beneath their feet, and have come out with accumulated fortunes. To those who were succesful, their luck seems little less than marvelous, but men are too busy to marvel long. The good fortune of others arouses them to renewed energies, seeking, climbing, struggling, they all toil along. At times they risk their all in a devil-may-care spirit and again they pinch their dollars until it seems as if they would break them. For them everything is at stake. This new Golconda offers to them another chance to gratify their ambi-

Photo by Allen

MAIN STREET, GOLDFIELD, JAN. 1904

tions, their hopes, their lives. They see the land beautiful in the distance and the vision incites them anew. Perhaps still further back of this fairy land of glittering and partly realized hopes, rise up the gaunt shadows of past failures or unsatisfied desires.

The barren plains and bald low-lying hills, rigged with derrick upon derrick, shaken with dynamite, rumbling with the whirl of the air-drill, is to them the field of the cloth of gold. The town is continually on the *qui vive*. From banker to waiter every nerve is at tension, every ear is awake to hear the first rumors of a new strike. Every home has its treasure box in which is laid aside the money with which to purchase in every new and promising investment, and every home has its hidden stock which is thumbed and fingered over and over, while the mind is dreaming air castles of the future.

The stories of Nevada's wealth that reach outside ears are listened to for a moment and then forgotten, but truth overtakes and passes fiction. However, the state of Nevada is not all gold. While its mines first equalled and then left far behind the production of the foremost gold camps of the world, yet much money has been spent in developing properties from which little ore has been taken.

In Goldfield as in former mining camps there is a great element of luck. The original lessees of the Hayes-Monette lease on the Mohawk spent over $100,000 in development and found nothing. They drifted toward every point of the compass excepting one. When Hayes and Monette took up the lease, there was only one thing to do. They set out to work in the only direction which had not been employed, and their drills bored into a body of ore which bids

Copyright by A. Allen

GOLDFIELD IN NOVEMBER, 1903

fair to surpass the days of the ancient Comstock. The richest of the Mohawk ore is a gray-black in appearance. This gold, for gold it is, is plastered over the face of the quartz for the thickness of one-tighth to one-quarter of an inch. It is so rich that the ore is stored in the vaults of banks for safety.

Similar to the Mohawk is the story of the Reilly lease, which produced over $1,000,000 in twelve weeks. This property had been abandoned by the original lessees, who had spent all they could rake and scrape of their own holdings, and then had borrowed from their friends in order to make a mine before giving up in despair. In this case the rich vein had been cut through and covered up by a treacherous miner who afterward sold his information, like Judas of old, for a pittance. Many another lease has without return absorbed the capital of the man who started in with brilliant hopes;

GOLDFIELD, NEVADA,

yet the wonderful richness of the desert state repays the persistent delver in its depth more often than it disappoints him.

Some of these ore bodies are deep and some are not; some are rich with ore, some barely assay, and it all lends an indescribable fascination to the game. It is a gamble, a guess. It bears all the fascination of a game of chance, but, contrary to gambling rules, in this case the percentage is in favor of the man who plays. The down-sinking shaft will strike a body of ore, which grows beneath the feet; it broadens out, it develops marvelous wealth. The miner, and his one or two partners, all of whom have been "grub-staked" perhaps, awaken some morning rich men. For the first time in their lives they have difficulty in counting their money. They stand stunned in self-contemplation, and then, according to the man, they do one of two things—they either go on a long spree and drink and gamble until they are reddish-black in the face, and spend all they make, or they become men of affluence, pointed out as the successful figures of a mining camp.

In the uncertainty of just what is beneath the ground lies the excitement, and excitement is contagious. The mining camps of Nevada today are on fire with enthusiasm. They seethe with earnest endeavor, they sweat with exertion. The prodigies of the camp are numerous. At Goldfield, from a fifteen-foot hole, the owners of the Kendall mine took $100,564. In Manhattan, the great Consoli-

NOVEMBER, 1906 Photo by Allen

dated mine lying to the north of Timber Hill gave up its wealth in sacks which were valued at many hundreds of dollars to the ton, while at Greenwater the Indian metal has paid to so surpassing an extent as to cause a stampede that promises to surpass any discovery of precious metal yet found. Such finds of mineral wealth in this flat desert country of sage-brush and sand are like a brilliant remark from the lips of some man whose heavy and stupid features had led you to think him a clod. You are startled, you look twice and then you sit up and take interest anew.

The recent discoveries of mineral wealth in Nevada are barely three years old and yet the state has already produced in its short life over thirty million dollars in gold. About five million was produced during the first year of the excitement beginning with the discovery of Tonopah, and during the last twelve months an average has been maintained of from $2,000,000 to $7,000,0000 every thirty days.

Notwithstanding this production, Nevada's territory has been but scratched. Its veins have been opened, but underneath the arteries have not yet been touched. The pioners of a mining camp are inevitably poor. They are men willing to bear privation on a chance. They are sometimes "staked," and again they scrape together what money they can, invest it in a team of mules and a wagon or in a pack-train of burros, victuals and a tarpaulin, and pull out from civiliztion to search their way along the low-lying bluff or through winding valleys, continually searching, tapping, examining. They are lonesome men, these prospectors—many of them educated, slow to speak, given to much self-communing underneath the stars.

With the first report of a find by a prospector, if the exhibited ore assays well, there is an immediate deviation of the stream of wanderers who trail in restless discontent up and down the land. Pell-mell they tumble into the new district. By auto, by stage, by every method of conveyance, some even on foot, they hurry toward the new discoveries. After them, treading in fact at their very heels, come first the saloon man and gambler and then the small merchant. Then, if the boom still increases and shows evidence of permanency, comes the promoter of active mind and ready tongue, the advance agent of the capitalist, and then the flood-gates are opened and in pour men of every walk of life.

The stories of the wonderful wealth of Nevada have gone east and been thrown back as the vaporings of a desert Munchausen, but they have returned again and again, backed by letters from Westerners with relatives in the East, from Easterners who had come West to see for themselves, from prospectors seeking capital for development, from engineers sent to investigate and report, from capitalists hurried to Nevada at the secret instance of trusted agents, until acting and re-acting, repeated over and over again, they have pried loose the equilibrium of Eastern investors and started down the slope of 1907 the most gigantic mining boom seen in the United States since the year 1881.

More than any one State and city in the Union, Nevada and Goldfield represent every type of humanity. Forty-eight ex-convicts are decent and law-abiding citizens within this busiest of mining camps. Twelve graduates of leading colleges of the land are acting as barkeepers or waiters or common miners; while many another college man, more fortunate, has become a mining expert, or is the legal adviser of some of the tangled disputes that arise within the camp. Bunco-steerers abound. The most artistic card-sharks in the world have been attracted by the glamour of the excitement and by the greed for the immense sums of ready money which accompany prosperity based upon mines. Mingled with these men are the rougher and simpler type of prospectors and plainsmen who have been bred into honesty and the singleness of purpose that goes with an isolated life where the sight of a fellow creature looming up on the horizon is a blessing undisguised.

Nevada is the stronghold of the union miner. The hordes of men who were driven from Colorado in the Sherman-Bell episode of several years ago have taken up their homes in the new camps and have so joined themselves together as to render themselves a power to be considered and dealt with in everything pertaining to the true interest of the new districts. They have not until lately shown the disposition of the old Cripple Creek days, and it is to be hoped that they never take an attitude such as made it necessary to drive their companions from Colorado.

Stock brokers abound. Their offices are as thick as saloons, and the latter number twenty-five to a block. Their immense mail matter floods the outgoing trains and overwhelms the railway mail clerks with the enormity of their business. No city of similar size in the world is today sending out one-half the mail matter that passes through the hands of the Goldfield postoffice clerks every thirty days. These employes of the Government are the hardest worked men and poorest paid of any who live in the camp on the desert. They obtain merely the usual scale of wages. which is prevalent in

Photo bv Allen

A HOLIDAY IN GOLDFIELD

other cities and the bitterness of it all came to the surface somewhat over a year ago in a strike which stopped the United States mail and crowded the room of the postmaster with a throng of excited brokers. Much credit is due the original postmaster of Goldfield, who, by the way, was a woman, Mrs. Ella Collins, and who paid for much additional help out of her own pocket in those early days when there were no private boxes in the postoffice, and when a line of impatient seekers of mail trailed through the front end of the office building and circled in a snake-like file half around the block.

Over eighty million dollars was invested in Nevada stocks during 1906. The fall of the year just closed, saw one of the most sensational rises in market value of stocks which has been known to the mining world. Led by the Mohawk mine, shares in which sold upon the streets for twenty-five and thirty cents only as far back as last April, and which climbed higher and higher until they attained the price of $20, stocks advanced rapidly until many which had been originally offered at twenty-five cents or less passed the dollar mark. A bootblack, whose mother had entrusted a certain wealthy mine operator with $300 of her son's savings, was made rich by investment in Mohawk stock. Men and many women were raised from poverty to wealth through long-forgotten purchases of stock to which many of them had long since ceased to attach any importance until recalled to their mind as a hazy memory of something that had occurred in the past.

On the strength of the great boom, many wild-cat companies were born and flourished for a time. Much stock was sold that possessed little value. Much money was obtained that was never invested in the ground, yet with it all, the tremendous richness of the district forbade anyone to disparage any property, for it is a mining-camp axiom that no one is able to see beneath the ground, and the most unlikely looking territory in this wonderful State is apt to prove a veritable treasure-box of ore. The effects of Eastern and California capital are plainly apparent in the camps of today. Where formerly a narrow-gauge road, crammed with passengers, puffed its dusty way toward Tonapah, and a stage swung and swayed through the hummocked plain toward Goldfield, today a Pullman car is brought directly into camp from the Oakland pier or is switched from the through Eastern train at Hazen and attached to the Goldfield sleeper. Contracts for railroads have been given to those who seemed the most capable of carrying through the work in the least possible time. Steel rails are lining the desert and lessening the hardships of travel and hastening the arrival of supplies and freight. The railroad south from Goldfield has nearly reached Rhyolite, while the railroad north from Searchlight, connecting with the Salt Lake road, has already reached the southern metropolis.

For the past years the Tonapah road, connecting with the Southern Pacific at Mina Junction, has paid Mr. Brock of Philadelphia and his associates a bigger profit than any railroad of similar length in the world, and has been the incentive that has started the same group of men toward the construction of the road from Tonapah to Manhattan to be completed within six months from date. The "Borax Smith road" (the Tonapah and Tidewater Ry.) in the south is being hastened toward the wonderful copper camp at Greenwater and Senator Clark is already hauling ties for his railroad to reach Greenwater from a junction a little south of Bullfrog.

Among the most sensational stories of rapidly acquired wealth is that of George Wingfield, the business partner of Senator Nixon, who climbed the rungs of the ladder from a cattle-man compelled to seek a loan from Nixon's bank at Winnemucca, to the owner of a gambling establishment at Tonapah, and finally to a business partnership with a United States Senator. The story goes that Mr. Wingfield, who is but twenty-nine years old, so impressed Nixon as president of the banking institution, that the Senator refused to accept a diamond solitaire ring as security for a loan, and gave the required money without note, upon the word of the penniless applicant that he would repay him at a later date. When Wingfield arrived in Tonapah he wrote again to President Nixon of the bank at Winnemucca, asking for more money, and the same was promptly

Photo by Dickerson

SHAFT ON MONTGOMERY-SHOSHONE MINE
Owned by Schwab and associates, and valued at $20,000,000

forwarded to him. Wingfield was a shrewd gambler, and his business in Tonapah prospered more and more as the days passed. He gained many friends among the rough men of the camp and gained much insight into the mining discoveries around about. He gradually obtained control of claim after claim, always "cutting in" his former benefactor, Mr. Nixon, for one-half, until gradually he obtained control of many of the best prospects in the state. From then on his influence grew. Backed by the strategy of the older men, he bought and sold and manipulated until his purchase, for the sum of $10,000,000, of four of the great mines of Nevada and their consolidation into one, made him the joint owner of eleven producing properties in that Golden State. As he became richer

A BANKER'S HOME

and richer, his money went into other fields and Wingfield the ex-rancher and ex-card-sharp became Wingfield the mine owner, the banker, the great promoter of business enterprises and a man whose influence is felt throughout the length and breadth of the State. The stories of his kindness to his former associates, of his generosity to his immediate friends and employees, are many and are typical of the largesse which seems to flow from the hands of these men, who are engaged in a pursuit in which dollars count as cents and thousands are reckoned even as hundreds are estimated in other vocations.

The startling question today in Nevada is, not the future, but the immediate now; the paramount individual question of gaining sufficient capital to develop the discoveries already made. It is not improbable that there are in the known gold district of Nevada today, a thousand claims that have the making, if not of a Mohawk, at least of great producing mines, and they remain unknown simply because the individual acquainted with the discovery is not equipped with sufficient capital to sink shafts and run the drifts necessary to convert the prospects into an organized producing mine.

It is safe to say that Nevada's great porphyritic belt, extending from the Comstock, with its record production, to Searchlight, and embracing an area four hundred miles from north to south and sixty to one hundred miles from east to west, is rapidly being proved the richest gold-bearing zone in the world. But while developments of the past four years have given to this great zone a score

IN RHYOLITE, NEVADA

or more of mining camps, of them all Goldfield has proved itself the richest.

While Tonapah thirty miles to the north is sending an output to the smelters averaging $40 per ton, and Bullfrog seventy miles south is making history with the production of ore having a tonnage value of $30, the Goldfield district, situated geographically in the center of the great Searchlight belt, is maintaining an output of ore having an average value per ton of from $300 to $800.

Compared with other records, nowhere in the world's history of mining can such a high average of values be found. The average value of production of Cripple Creek was $30 per ton. In the Black Hills the mines are outputting ore of an average valuation of less than $5.00. The general value of ore mined in California ranges from $5.00 to $20.00 per ton. Altogether, the African production of gold in the Transvaal shows an average tonnage value of $17.00 per ton. This feature of Nevada's production, has had and will continue to have an important bearing on the part this wonderful State is destined to play in the world's output of gold. In the history of mining, it has almost invariably been found that exceptionally rich ore veins possess little, if any, permanency. The rich ledges of ore that have been encountered have proved "pockety." This was proven true of Cripple Creek, where practically all the exceptionally rich ore has been found in narrow streaks and stringers. Stratton's great Independence mine has produced some phenomenally rich ore, but its average output has not attained a valuation above $40.00 per ton. The Portland's output has not even attained this average, while the El Paso property, which heretofore has produced the highest grade of ore in Colorado's great gold-camp, has yielded a tonnage having an average value of but $55.00.

While there were many who were emphatic in their declaration that the ore-bodies would not go down, all of these theories have been proven fallacious. In the great Florence property, ore has grown richer with depth, as has been the case in the Combination mine and in the lower workings of the Jumbo, the Red Top and others. The gray-black ore of the Mohawk is the richest of them all, while the peculiar formation of the ore and the richness of the value is a surprise and a revelation to the miner and the mining engineer alike. The occurrence of gray copper in the ores of the Florence, January, and other prominent properties is considered an indication of the permanency of the ore bodies. After Tonapah had been discovered by Jim Butler and his recalcitrant burro, the belief in the richness of the entire district extending south from Tonapah was firmly implanted in the mind of the "desert rats," who made their living first by prospecting, and then by turning over their prospects for enough capital to provide them food for a month and a day.

But Al Meyers, who recently sold a hundred thousand shares of Mohawk stock for $4.00 per share and then watched it climb to the dizzy height of $20.00, was the first to set actively to work upon his belief. In the fall of 1902 he started out on a prospecting trip which took him into the Goldfield district, then entirely unexplored, although a train had passed over many of its richest mines for years past. In 1903 he had located the Combination Group. In passing over this ground one day he had kicked loose from an outcropping, a block of quartz. This he had assayed and found that it averaged $4.00 in gold per ton. A piece obtained a foot lower gave returns of $22.50 per ton. From this assay Meyers concluded that further development would demonstrate that the values increased in geometrical ratio to the increase of depth. This theory proved correct. In performing the ten feet of assessment work required by law to hold the title to the ground, Meyers opened up th Combination group, which has since proven to be one of the richest mines in the world, and was recently purchased by Nixon and Wingfield. He was without money to develop his find at that time, however, and so, having secured the property by correct notices and sufficient assessment work, he passed on to other ground and located the Mohawk and many additional claims. In October he bonded the Combination Group to capitalists from Chicago for

$75,000. Five thousand dollars was all that the purchasers were obliged to put up from their own pockets. The mine paid for itself from the grass-roots, and the enterprising financiers from Chicago completed the final purchase price on their property from the property itself.

With the completion of this sale and the remarkable discovery of ore which immediately ensued, began the actual commencement of the Goldfield excitement, which has increased in geometrical ratio until today the district is the scene of the most aggressive and tremendous mining operations ever witnessed in the world, while a pay-roll of some 3,000 men and an output ranging from three to seven million dollars a month, has been the record for the last four

ONE METHOD OF STORING ORE

months. In the short space of two years, a town of 15,000 people has been laid out on substantial lines, streets lighted by electricity, telephone and telegraph connection established to the great cities of the coasts, and complete water and sewerage system has been installed. Goldfield's position as a distributing point for the entire mining district, including Manhattan, Silver Bow, Goldcenter, Kawich, Wonder, Tokop, McGibbons, Tule Cañon, Bullfrog, Lida, Rhyolite and Silverpeak, has been firmly established.

Nevada will remain a boom State for years to come. No sooner does the excitement wrought up by a discovery in one locality steady itself into earnest development along industrial lines than another discovery, even more wonderful than those preceding,

BULLFROG,

sweeps miner, prospctor, and tenderfoot alike into a renewed frenzy of pursuit.

In such manner the news was borne into Goldfield that seventy miles to the south, tremendous ledges of medium-grade ore had been found, that reared their heads for miles across the scraggy hills, and would prove great store-houses of mineral wealth which could be worked by the children of the present generation and by their unborn children yet to come. At first the original Bullfrog, with its beautiful blue quartz, was the center of the attention which swept in a widening circle to the south. Town after town sprung to life and dwindled to nothingness as new discovery after new discovery lured the miner still further on and on. But finally the center of the magic circle was reached, and Rhyolite, having captured the wandering population, became the central city of the southern desert.

To Messrs. Clark, Benson and Ladd, belong the credit of the founding of Bullfrog. It was their money that staked the persistent prospectors who, travelling on and on, enduring hardships almost unendurable, cutting the tops from their boots and strapping them to their feet when the original boot-soles were born through, carrying water for miles on rounded shoulders, bending over every promising ledge, keeping continually at it until they finally located the district which "looked good enough to them." Then their "grub-stakers" came to the front and capital poured into the district to develop it.

NEVADA

While the excitement around Ladd Mountain and Bonanza Hill was at its height, Robert Montgomery, familiarly known as "Bob," walked into camp one evening and offered for sale certain claims, which he stated bore promising indications of ore. When asked the location of his property, he informed his hearers that it was about a mile up the cañon, pointing with his thumb over his shoulder to the east. Not one of those listening to his siren song responded; in fact, they gave Robert the laugh and he was forced to retain the unwilling possession of his ground. It was but eighteen months afterwards that he bonded the now famous Montgomery-Shoshone mine to Chas. M. Schwab for five million dollars and the property which he had so anxiously sought to dispose of for a paltry thousand, placed him and his family in the ranks of millionaires.

Rhyolite and its surrounding mines teem today with activity. The advent of the railroad from Searchlight on the south has lightened the effort of development to an appreciable extent. The town is booming. The miners dot the neighboring hills and the activity is second only to Goldfield. There are no sensational discoveries of high-grade ore in Bullfrog, but there are tremendous deposits of ore and gigantic true-fissure veins that maintain a steady average of value and which will pay a steady income for development for years and years to come.

First Tonapah, next Goldfield, then Bullfrog, after that Manhattan. What a similarity of legend, which traced down still seems to be the truth, surrounds the discovery of Nevada's mining comps.

In the case of Manhattan, the story goes that Jack Humphreys, tippling quietly by himself as he rode in a jovial mood down the trail from Austin toward Tonapah, alighted from his horse to sleep off, beneath the shade of a pine tree, the effects of liquid he had imbibed; for Manhattan, may it be known, possesses more timber than the balance of Nevada's camps added all together. When Mr. Humphreys awoke from his sleep, his eye caught the setting sun and with a hasty epithet he scrambled hurriedly to his feet. His heavy nailed boot-heel slipped on a point of rock and it broke away beneath his tread. Upon examination, the character of the ore attracted his attention, for the eye of the miner is as keen for gold as the eye of the old-time Indian was keen upon the trail. He examined it and his sudden hurry to be off was changed in the twinkling of an eye. The miner's magnifying glass was whipped from his pocket and beneath the lens, in the rock within his hand,

Photo by Dickerson

THE MAIN STREET OF RHYOLITE

were many specks of free gold. His find was made on the first day of April and his claim was named the April Fool claim. He finished his journey, but never forgot his find, and later returned with three stalwart brothers and his father, when the entire family busied themselves in the location of what seemed to them to be the best claims of the district.

The Humphreys knew Lawyer Vermilyea of Goldfield and interested him and several newspaper men. The camp was thoroughly investigated and the results were so promising that the news was speedily given to the world and a town three miles long by a thousand feet wide grew up over night in the narrow valley of Manhattan. Further investigation merely intensified the value of the discovery until today Manhattan possesses thirty working properties and is producing a tremendous output of high-grade milling ore. Its most recent discovery has been a wonderful find of high grade copper ore (and it is the opinion of mining experts that the entire Goldfield, Tonapah and Manhattan territory will at a depth of 500 feet show a great underlying blanket of copper.) The discovery was made on the Manhattan Dividend property on Timber Hill and assays the wonderful average of 14.5 per cent. copper, with cross veins which give values as high as $72.00 in gold. This mine is being developed assiduously and every foot of depth obtained is subjected to the interested scrutiny of mining men.

After the finding of Manhattan came the steadily growing belief in the tremendous copper deposits at Ely. Money was diverted into this camp for extensive development and today the entire mining

Photo by Dickerson

RHYOLITE, NEVADA, GOLDEN SCEPTER MINE IN FOREGROUND

world has been shown that Ely posseses copper in tremendous bodies and of splendid average grade

What may turn out to be the greatest discovery of the Nevada ore belt, as well as the latest one, is the find of copper at Greenwater camp, really located in California, but only a few miles over the Nevada border line, and considered by all mining men as a Nevada discovery, inasmuch as there is no ingress or egress except through the lower towns of Nevada and inasmuch also as Greenwater lies in the very center of the great ore belt that sweeps the entire length of the desert state, trending from the north in a slightly curved line toward the southwest through the lower part of California, then bending to the east and crossing into Arizona.

Grenwater has seen fifteen million dollars invested in its properties within the past six months. It has been the heated subject of discussion among the entire mining fraternity, and many noted experts have crossed the desert and endured the hardships of travel to this lonesome locality at the head of Death Valley for the purpose of examining its marvelous ores.

Donald B. Gillies, the mining expert of Chas. M. Schwab and the general manager of his Nevada property, has made the statement that he would readily transfer his entire holdings in Nevada, and he possesses many splendid ones, to spend the proceeds in the development of Greenwater, a camp which he proclaims to be the greatest discovery of copper ever made in the history of the world.

Peculiar turns of fortune stand out in the foreground of Goldfield's kaleidoscope mass of excited humanity, with the distinctness of an etching. The lawyer of whom it was said that he remained

Photo by Dickerson

IN THE SUBURBS OF A NEVADA MINING TOWN

in Goldfield merely because of his inability to dispose of his typewriter, now retains it and draws a check for $50,000 besides, with ease.

Meyers and Murphy, the original discoverers of the Golden Goose, today, by including all their holdings, possess a bank roll that is close to one million. Henry Weber, formerly a hotel clerk in Denver, is today erecting in Goldfield a five-story building of stone and reckons his cash assets at beyond the half-million mark. Butler, Wingfield, Nixon, Parker of the Colorado & Southern, Lockhart whom he staked for twenty-five years before obtaining any return, Loftus also of Denver, Jim Davis, L. L. Patrick, Milton Detch, and more than one hundred and fifty others, as a result of their unfaltering belief, restless energy and determined effort, are ranked over the six figure mark in this world's goods.

These men, almost to a man, less than thirty months ago were unable to spend $30 for a suit of clothes without much mental ransacking in order to determine the wherewithal with which to pay the bill. But the pendulum has been swinging and has certainly been swinging abnormally for an army of successful men whose wealth has come to them, not from the avenues of commerce, but from the vaults of the hills.

Because the immense total of nearly one hundred million dollars has been diverted from the channels of Eastern investment, there has been much scathing criticism of late of Nevada and its wonderful mining camps. A prejudiced press has in several instances distorted the truth and repressed facts. Nevada today is the richest mineral bearing State in the world. Discovery after discovery, mine upon mine, ore body after ore body, extending for miles on

Photo by Dickerson

AN AVERAGE PROSPECT HOLE

the surface and for miles beneath, open up tremendous possibilities for those who are willing to engage in the strenuous life, where every waking moment is filled with fiercest struggle, and every hour of sleep away from the forge of possible discoveries is taken with reluctance and looked upon as a hindrance in the terrific game that is being played in the mad struggle for wealth.

The youngest and most recent of the known mining discoveries, Nevada easily outstrips them all. The great State is full of anecdotes, its history is crammed with picturesque incidents which are daily added to from a wealth of material which is ever new and always exciting. The panorama is a marvelous one. Individual types such as are seen nowhere else in the world abound. From the lowest to the highest, from the most selfish to prodigals of generosity, from the most stupid to the most intelligent, Nevada's civilization runs the gamut of personality and abounds in human interest.

Los Angeles.

GOLDFIELD AND THE GOLDFIELDERS

BY ONE OF THEM

IN THE course of American progress it becomes necessary for a portion of our people to take upon themselves the burden of the development of its remote districts, and bring to our country the advantages of their hidden wealth.

"We, the signers hereto, have undertaken this self-imposed task with every confidence that our efforts will prove of great benefit to the whole people, and that the mining operations of Nevada will add greatly to our National wealth and great gains to those who shall lend financial aid.

"Here, in the desert, we have builded cities in permanent form; here we have brought machinery, men and money, that we ourselves may be enriched and that we may secure for the uses of mankind the elemental treasure so bountifully bestowed by nature.

"Goldfield, in the desert, is a city three years old, with a population of 15,000; with churches and schools; with four prosperous banks; with many paying mines and many more under development; with costly reduction works for the immediate treatment of ores; with a stock exchange where prices are fixed by men who have intimate knowledge of values; with its affairs in the hands of educated men and women.

"Here is the greatest proven gold-mining district in the world, where operators and miners find prosperity and peace in an endeavor for common betterment.

"Here we are located in the state of the smallest population, with wealth sufficient for an entire nation. Here, we have sought the aid of the whole people, that benefits for the whole people might result. Here, have we prospered and given back to the development of Nevada the gains of our prosperity.

"We began with prospectors and burros, and in a few years have builded cities, begun an unprecedented railroad construction, and are established on firm foundations for all time.

"The possibilities of our Desert Country have not been unduly magnified, but are vaster than our utmost imaginings, and we point with pride to the appreciation which has marked our progress, growing daily in magnitude wheresoever our fame has spread.

"We of Goldfield, who have sown our faith, have garnered golden harvest, in which we invite all people to participate."

This proclamation, followed by the autograph signatures of a large number of the representative men of Goldfield, appeared in December, 1906, simultaneously in the newspapers of thirty cities of the United States.

Beneath the siren song of the press agent, do you not catch a deeper note prolonged from another Declaration of Independence? Flamboyant perhaps, but full of the exultant pride of achievement and forecasting greater worlds to conquer.

If you would survey the monuments of these pioneers, from the foothills of Columbia Mountain, look around! Here, not many months ago, stretched an unwatered and unshriven desert shut in by shrunken and uncanny hills, metes and bounds of a moribund world. An amphitheatre of silence, unbroken since those hours in the twilight of the earth when yonder volcano last volleyed its subterranean thunders. What was once a dust-puff, whirled from the trail of a mountain wolf, has grown into a cloud which now and again covers the landscape like a pall. Out of it emerges the twenty-

Photo by Allen

"MOHAWK'S GOLDEN ACRE," GOLDFIELD

mule team, advance guard of the railroad, hurrying southward to Death Valley. The malapi is vocal with cries of the argonauts and their camp followers. Their tents and housings rise over night like an exhaltation; their chariots shod with air go to and fro like weavers' shuttles. Out of the volcanic tufa at our feet, from Mohawk's golden acre, spring a forest of gallows-frames. Their timbers stand as thick as derricks in an oild field. Their engines are never silent. Day and night the sibilant cough of the exhaust marks the rising flood of Pactolus. Mohawk's acre holds the world's record of gold production. Thirty-five dollars a minute, two thousand an hour, fifty thousand a day, and in the last one hundred days of 1906 five millions of gold is added to the exchequer of mankind. This fabulous vein, purse of Fortunatus, is cached in a Monte Cristo cave, one hundred and fifty feet as far as explored during the golden age of the leasers.

Figures convey a meagre and misleading idea of the contents of this treasure vault. To say that 45,260 tons of ore extracted under the Mohawk leases produced $5,228,616 of gold in less than six months falls far short of the fabulous reality. There is still nearly $3,000,000 of high-grade ore to be added before the accounts of the leases can be made up.

At the lower levels the nether Gods seemed to have refined the mass by volcanic fires seven times heated. They poured the previous metal into fissures and set them in immense walls one hundred feet in length. Streaks and stringers of quartz widened into seams, and these into ledges.

Copyright by A. Allen

THE GOLDFIELD STAGE OF DAYS OF NOT LONG AGO

I have seen the Goldfield miners, candles in mouth, clinging like bats to these walls. Under the wavering light every eye gleamed with avarice. The sharp end of the pick pried the rotten quartz from its matrix, and crumbled the almost pure gold into the hollowed hand, where it sparkled like fireflies in the dark.

This dust, too precious to be mixed with the coarser rock, was packed into little sacks, which as fast as they were filled were hid in near-by chambers in the stopes. Stout doors secured this garnered wealth, and watchmen guarded it night and day.

As the leasers' time dwindled to weeks, then days and finally hours, miners worked shoulder to shoulder in four-hour shifts, relieving one another that not a moment should be lost.

Places in the ranks of this army commanded a premium to their fortunate pre-emptors. It was like the waiting line before a box office anticipating some noted dramatic spectacle. The right of seizure and search had not yet been placed on the statute books of the underground world of Goldfield. Therefore, under their shirts the men hung cunningly devised harnesses, taking tithes of all they extracted under the very eyes of their employers. It was better to suffer this loss than to replace the knave with a more honest and less skillful workman.

Deadly perils lurked in these lofty stopes, imperfectly supported by timbers. Every blast loosened more than the debris of the drifts. The earth groaned and travailed, and the walls and ceilings of the stopes threatened collapse and wholesale extinction of life. Yet such is the lust for gold, that the waiting line at the shaft watched

Photo by Dickerson

AUTOMOBILE STAGE BETWEEN RHYOLITE AND BEATTY

enviously the workmen descend, knowing that the privilege of a few hours' labor might mean a month's income under ordinary conditions.

Is this a new tale of Scheherezade, translated from the Arabian Nights, or a modern miracle-play of the desert with stage miners, heroes and villains playing their brief parts in the limelight? Thus many men at a distance thought when these stories began to circulate around the world.

But those of us who were a part of these scenes can still hear the ceaseless rumble of mortars pounding ore in those pseudo assay-offices which honeycomb the back streets of Goldfield. Scores of their proprietors have grown rich by buying stolen ore piecemeal that ran one hundred thousand dollars to the ton.

We can recall the wagon, heaped up as with sacks of grain, three hundred and fifty thousand dollars to the load; and that forty tons of high grade, a carload lot, valued at one million dollars, was a very real and tangible asset of the Hayes-Monnette lease.

Surely nowhere on earth has the God of Luck played so many pranks, or assumed so many protean shapes, as under the shadow of Columbia Mountain.

"Right under that big hill lies my fortune," cried A. D. Myers, the prospector, to his partner, as they lay in camp one day in the fall of 1903 on the eastern slope of the Tonopah range. The next morning he went forth unerringly, as one led by fate and found the Mohawk. The rainbow's pot of gold became his by right of location. In one day he set his stakes over ground rich enough to pay the na-

Photo by Allen

A GOLDFIELD GALA DAY

tional debts of half the petty principalities of Europe. His original prophecy came true, but if the vision had been clearer he might have stood on the Mohawk outcropping, a modern Monte Cristo and cried, "The world is mine," for under his feet lay the riches of Ormus and the Ind.

As well prophesy where and when the flake of snow will fall from some passing cloud, as to forecast the gifts of fortune in a mining camp.

"From where do you think the gold comes?" asked the first governor of Victoria of an old Cornish miner in Australia. The miner replied, leaning upon his pick and shaking his head lugubriously, "Where it is, there it is, and where it ain't, there be I."

The history of the Combination Mine, a premier location of Myers, is replete with dramatic surprises. This mine lies over against the Mohawk, a little gulch separating the two claims.

When a prospector has appropriated as large a part of the earth's surface as he can measure off in a day's work, he rests from his labors and sets forth to find a purchaser. If he has been fortunate enough to strike paying ore at grass-roots, he feels that a fortune is within his grasp.

Seventy-five thousand dollars looked like a king's ransom to the original group which located the Combination ground, not much more than an open cut on the crest of a hill, with quartz that assayed $28.00 to the ton. It might prove no more than a surface deposit; so, like many others, they dropped their picks in the trench, fearing

Photo by Allen

A BUSINESS CORNER IN GOLDFIELD

their fate too much to risk all in sinking further on the vein. Others who had seen the same showing were more adventurous. A Colorado man, L. L. Patrick by name, held the option on which the first payment of five thousand dollars was due and payable. Only twenty-four hours remained on Patrick's option and the money from his eastern associates had not arrived. A Tonopah mining operator and a Winnemucca banker had agreed with the owners to advance this payment and take over the property at the same price on the expiration of Patrick's option. Here was a situation replete with dramatic possibilities. The anxious owners already beginning to spend their immense windfall; the holder of the original option watching time pass on the dial and fortune slipping through his fingers, with each tick of the clock. In the background the second comers were waiting and watching, confident that the coveted property would soon fall into their hands. Imagine the importunate messages passing out of the desert, each more despairing than the other, and into the offices of the Chicago capitalists, to whom such far-off issues seemed trivial, yet not to be hastily decided!

Under the sage brush slept The Treasure. It alone was serene, quiescent and indifferent. If it had been awakened and endowed with the gift of sight and speech, The Treasure might have overlooked the contention and selected its own favorites, and thus diverted into new channels the inevitable flow of its golden stream. It was the prize of nature and of time, and the players recked not of it save as a counter in their game. The die falls with the click of the telegraph key; Chicago "has deposited five thousand dollars to the credit of L. L. Patrick to pay the first installment on the purchase of the Combination Mine." It lacked a few minutes of three o'clock in the banking house at Tonopah when this message was received. Perhaps it was The Treasure that spoke after all—who can tell? On what slight motives and impulses hang the greatest of human decisions.

The other chapters in the history of the Combination Mine follow swift and fast. Almost the next blast in the shaft discloses bonanza ore, and before the second payment was due the entire purchase price of the mine was taken out of the original workings. Boston and Chicago joined hands to develop this great property. The crest of the hill was covered with a concentrating plant and a stamp-mill of the most improved construction. Across the desert pipes were laid to a boiling spring, its waters were brought to the mill, and proved more than sufficient for its needs.

Visitors to the Combination mine are taken to the site of the original discovery and told the story we have narrated. It seems more effective as one looks down into the opening, like the mouth of a crater—"Glory Hole" as it is called—where the workmen can be seen far below shoveling the broken quartz into the stopes, from whence it is lifted by the neighboring shaft, and automatically fed under the stamps and through the chlorination tanks.

This mine, after enriching Patrick and the eastern shareholders to the extent in dividends of nearly a million dollars, has finally been joined to its neighbor across the little gulch. The Treasure has changed owners for a price, fabulous at least as compared with the mess of pottage for which the original prospectors bartered it. But with the change of ownership passes away the danger of apex suits, that bane of mining camps; and if The Treasure, a little depleted after two years, is glad or sorry, who knows? Certainly not the writer of this chronicle.

Mercurial Goldfield lies almost at the bottom of the railroad tube. Its prosperity fluctuates according to traffic conditions. At the top of the tube lies Reno, well watered, housed and comfortable. At the bottom yawns Death Valley full of borax, copper and dead men's bones. If the Lord of the Railway sends us cars, we can fill them with gold and after many days, they should return to us laden with coal, ice and edibles But if he forgets us, we freeze in winter and go parched and hungry through the long summer days. This is a condition, not a theory.

During Nevada's interregnum between Comstock and Tonapah, these desert railways dwindled into dust-streaks. Stockholders went without dividends and bond-holders without interest. An

Photo by Allen

THE GOLDFIELD FIRE OF SEPT. 1907

occasional train crawled in those days as far as Mina from the main line of the Southern Pacific, and beat back slowly against wind and snow-storms that threatened shipwreck at every mountain gorge. Discovery of the Mizpah mine by Jim Butler, on the present site of Tonopah, reversed all the old conditions. The profits of the Mizpah mine extended the narrow-gauge from Mina to Tonopah, and subsequently to Goldfield. Mina is a Turkish word, which being interpreted means, "Gateway to the Valley of Promise." Fortunately for the railroad builders, these remote valleys soon began to redeem every promise a thousand fold.

It is good magazine form to abuse the transportation companies, but these early invaders of the desert had their troubles as well as their problems. They undoubtedly minimized the importance of this new territory. They delayed standardizing the old narrow-gauge track and providing adequate equipment and sufficient sidings. Their rates were all the traffic would bear. The Goldfield-Tonapah road at the present time about pays its cost in dividends to its stockholders.

An interview with the managers in Philadelphia discloses the fact that they are very human. They are mine owners as well as railroad men. They know the needs of the community. Railroads, they say, like mines, are not operated for the health of their owners.

It is true that during the winter months it took ten days to get a ton of coal from Oakland to Goldfield, and the freight charges were $10.00 per ton, or $400 per care. This for an ordinary 24-hours'

Photo by Alexander

A PROSPECTOR

run. But "responsibility was not ours alone," they allege. "The Southern Pacific system, like our own connecting road, was short of cars, locomotives and fuel. An unexpected influx of population into southern Nevada made such demands upon railroad service that the most we could do was to carry ore out and provisions in. Good railroad men were hard to obtain. They turned miners and speculators and abandoned the locomotives on the track to plunge into the Walker Indian Reservation and prospect for gold. When overworked locomotives broke down under continuous service, we could not obtain new ones from the eastern builders. They refused to accept our orders except for six months delivery."

Thus runs the railroad managers' tale. Underneath it lies one undeniable fact: The Railroad KiKngs, to whom these Western provinces are merely places for their pawns, their knights and their rooks in the railroad game, were, in 1906, too intent upon their moves in the Northwest to provide for the needs of the desert.

But the cry of the Mohawk acre is more insistent than that of all other ungathered harvests anywhere. It will not be denied. It finds an echo in every banking house where gold is weighed in the balance and found wanting.

Hence these rumors of new railroad projects which now fill the air. King Harriman, it is said, looks covetously at this Naboth's vineyard. He would like to add to the Southern Pacific system this Mizpah tentacle which is slowly feeling its way past the tents of Goldfield towards Bullfrog and the Greenwater country. It is annoying to rule the body, and not one of its little members like the Goldfield-Tonopah road.

Photo by Dickerson

ON THE WAY TO THE PROSPECT

Hither also comes with his engineering and surveying corps, Gould, by way of Salt Lake and the Western Pacific. There are fertile valleys and rich mining camps between Winnemucca and Tonopah. From the south a feeder of the Salt Lake and Los Angeles road has travelled northward as far as Beatty. By early spring it should be bringing goods in from Los Angeles and dividing the ore shipments to the Salt Lake smelter with the Southern Pacific.

This pursuit of the people who dwell in tents is in amusing contrast with the autocratic temper of the managers a few years ago in this country. In those days Hawthorne, the present county seat of Esmeralda County, boasted of a branch railroad. The railway officials decreed a new townsite some miles removed from the one established by the first settlers. The settlers naturally expected a road to come to them. Its managers, well supplied with cheap town-lots, refused to meet the old townspeople even half way. "If you don't move up to us," the railroad men said, "we will pull up our toy tracks and water-tanks and leave you stranded as before in the desert." "Pull away," replied the population of the county seat. The railroad took the town at its word, folded its rails and left the lawyers and judges who are obliged to visit the county seat during court terms to travel there over the abandoned road-bed in a ramshackle stage—which they do to this day. Let it be noted that the State of Nevada has not yet provided itself with a railroad commission, that modern weapon against railroad aggression.

It must have been a remnant of the same feudal spirit which induced the Tonapah road to plant its station a mile and a half from

Copyright by A. Allen

A USEFUL FRIEND TO THE PROSPECTOR

the town of Goldfield, expecting the town to come to it. This it implacably declined to do, albeit with much profane increment every train-time when compelled to pay two bits to the bus drivers and endure the added pangs of a corduroy road.

When a railroad commission is appointed, one of the first prayers of the camp will go up for an improved train service between Goldfield and northern points. It should be possible for the traveller to arrive and depart by night trains, thus mitigating the asperities of the long ride through the desert—in summer hot, dusty and uncomfortable, and in winter bleak and depressing. Under such improved conditions, however, he will miss the arrival at night, the spectacle of innumerable dancing lights around the shaft houses of Milltown, the mad race of hacks and busses through the darkness and the sudden transition from the silence and calm of the outside world to the light and tumult of Goldfield streets

> "It is day all day in the daytime
> And there is no night in Creede."

Cy Warman's couplet applied to the Colorado mining camp is equally true of Goldfield. In the dominion of chance the sun never sets. The faro and stud-poker dealer, the croupier and crap thrower, work like miners in eight hour shifts. To the peripatetic philosopher, games as they are played at Goldfield and the mine, offer some analogies. Superficially, the action and atmosphere of the temple of chance and the Bourse are similar. Reckless and restless votaries gather in both to watch the fall of the dice or the

Photo by Dickerson

A PROSPECTOR AT HOME

rise of quotations. In the Bourse the caller cries out the bids of the brokers who press about his rostrum. A secretary records the bets, that is, the varying prices of the mining shares that are quoted on the board. Shut out by a low railing the excited speculators jostle and crowd each other.

During the evening sessions the fair sex is much in evidence. The seats reserved for the ladies are always filled. The wife of a mine owner who has made $5000 over night by a leap in Bullfrog Consolidated exchanges views on the market with the daughter of a miner. The father has just come off the night shift and brings tales of some new strike in the Diamondfield District. The former swings a line of 10,000 or 20,000 shares, the latter is satisfied with the buyer's option of 100 shares. Salaries of $150 a month attract stenographers from every state in the Union, and young and old alike speculate with absorbed ardor and varying fortunes.

"A mad scene, my masters," this gambling pit in the midst of a desert!

"What am I bid?"

"What is the stock offered at?"

"A thousand Triangle at 10."

"Sold!"

"I'll give 11 for any part of 10,000."

Such are the cries that startle the night.

When the market soars, lifted by orders from all parts of the world, this pit is a pandemonium. The crowd fills the room and overflows on the sidewalk, where a fringe of speculators kneel on

Photo by Allen

AN UNEXPECTED USE OF BARRELS

the iron grating and shout their bids through the open windows into the basement below.

Out of this underground room runs a telegraph wire which keeps the night session of the San Francisco Mining Exchange in constant touch with the fluctuations of the local market.

By the time this article is in print a direct wire to New York will have been added, involving arbitrage transactions with the "curb" and the mining exchanges of the principal cities of the country. When to the surplus money of the factory, shop and field are added the funds of capitalists, and investors in every part of the country are eager to trade in Nevada mining shares, it is easy to understand why brokers grow rich, and one stock exchange is not enough for Goldfield.

After the Northern Pacific corner in 1901, when that stock jumped from $100 to $1,000 per share, the air was full of tales of fabulous fortunes won and lost over night. He was lucky and exceptional who bought at the bottom and sold at the top. The par value of the railroad stock is $100 a share, and that of the Goldfield-Mohawk $1 a share. The present market value of the mine represents, therefore, an equivalent of $2000 a share in Northern Pacific stock, and the speculator has always been able to realize on his original investment from one hundred to one thousand per cent profit. Small wonder, then, that the average human animal, who takes to gambling as inevitably as ducks to a mill-pond, should be even now searching the mining stock reports for another Mohawk.

Between the Scylla of the pit and the Charybdis of the gaming

Photo by Allen

10,000 BOTTLES WENT TO THE MAKING OF THIS HOME

table, the average miner, however fortified by union cards, has small chance to save either the proceeds of his eight-hour-a-day labor or the rewards of the most strenuous "high grading."

In the old Comstock days the fall of the market was carefully planned to the end that the high wages paid should regularly return to the mine owners and stock manipulators, leaving to the mine worker at the end of each year little besides hopes and regrets.

History may repeat itself in Goldfield, but thus far miners, as well as all other classes of operators, have profited greatly by the rise in mining shares. So potent are the subterranean attractions, so irresistible the golden lure of stope and cross-cut, that men who sit for hours in the poisonous atmosphere of the saloons dealing faro will leave the game, not for rest or dissipation, but to go on a day-shift with pick and drill in some ground which they have leased on royalty from the owner.

Fashion, if you can, the dreams and imaginings of such men. With what feelings must they regard the blind worshipers of chance who crowd around their tables nightly, flinging away by a turn of a card or the wheel the earnings of hours of toil.

Only that is magnificent which is unknown. Neither miner nor croupier recks of the percentage against every great coup. Each is willing to play the other's game—and both lose in the long run.

If the Unions devoted a tithe of the effort they now expend in increasing wages and shortening hours, to schemes for improving the conditions under which their members live and especially enforcing habits of saving and thrift instead of reckless prodigality, they would show to the world larger reason for existence. But how tame, dull and uninteresting would Goldfield, the premier gold camp of the world, become by comparison with its present apparelling.

In long, narrow and low-roofed saloons, glasses unceasingly clink over the sweating bar. The man from the Rand stand cheek by jowl with the exile from the Yukon, Australia and the Argentines. He who was deported from Cripple Creek recites his wrongs to his sympathizers in typsy heroic style. If—too common mischance—the blind goddess frowns upon the gamester, the house hands out a bar-ticket and hope flows again under the spell of renewed potations. If money is lacking for lodgings, the saloon offers a refuge. In the early morning hours far corners are crowded with rows of prostrate revellers to whom a sawdust bed brings brief oblivion. Here shows the seamy side of a great mining camp, since for these on the morrow the Morgue may be the next resting place.

Conditions which existed during the fall and winter of 1906 are not likely to be repeated during the coming year, and this despite an impending invasion of newcomers from all quarters of the world. The winter was unusually severe and inclement; snow lay in drifts for weeks, instead of yielding to the first rays of the warm Nevada sun. Coal and wood sold at prohibitive prices, and no man's outbuildings were secure against his neighbor's imperative demands for fuel.

To add to the general suffering, the daily train, which should have arrived in the evening, was delayed for hours between Mina and Tonopah by freight wrecks and misplaced switches.

After spending the night in the crowded cars, the arriving passenger was unable to secure lodgings at any price. Seats around a hotel stove brought the price of a room and bath at the Waldorf-Astoria. The only pretentious hotel was totally destroyed by fire with unfortunate loss of life.

It was a season of unparalleled excitement on the Exchange and feverish activity among the leasers, those especially who held the "open sesame" to the Mohawk Treasure House, and whose opportunities of profit waned with the declining year. Orders for the purchase of stock poured in from so many quarters and in such amounts that the overworked and exhausted clerical forces of the brokers were physically unable to post and balance their books. The Stock Exchange was closed for three days. Exchanges have been closed in great trading centers, but generally for reasons other than the inability of their members to catch up with their business.

Such was the plight of Goldfield when emissaries of the Eastern press flocked down upon her. They came avowedly to spy out the land, to expose its pretensions and let the truth and sunshine of the outside world into its darkest corners.

Photo by Dickerson

BUCKING THE TIGER

A careful digest of the many articles intended for Eastern circulation leaves this impression upon the reader: "Goldfield is a bad place to live in and an easy place to die in; it is full of wicked mining brokers, who advertise worthless prospects as mines, and thus separate the innocent and unwary from their money at great expense of advertising in the very papers whose editors print daily warnings against mining shares and tips on the New York Stock Exchange."

On the score of accommodations or luxuries of living, this city of the desert, as we have shown, leaves much to be desired. Her critics forget that she is yet an infant less than three years old. In their swaddling clothes, and at such a tender age, even the great cities of the Atlantic Coast were without hostelries supplied with men roamed about their unlighted streets at night, and even now are not wholly exterminated from State, Chestnut and Wall streets. The chief difference between Lawson of Boston and McKenzie of Goldfield is that the latter has taken more money out of the ground than the former, although they are both in the mining and stock-brokerage business.

You may say that "good wine needs no bush," but McKenzie carried that *bonne bouche*, the Francis-Mohawk Lease, from door to door in Goldfield and Tonopah before he went to the public and secured the capital to sink his shaft on the Mohawk vein. Those who read his early advertisements and subscribed for his stock at twenty cents on the dollar are now able to secure a profit of 500 per cent on their investments.

Photo by Dickerson

A ROCK DRILLING CONTEST

In these articles we read a great deal about the shady antecedents of the heads of certain so-called financial institutions and their mining properties in various Nevada camps, but we are given no statistics of the output of the Goldfield mines, nor any information on which we can base a conclusion as to their merit and productiveness. In fact, these traveling correspondents, who came into the camp on one train and left on the next and spent the intervening time in collecting and tabulating stale gossip and hearsay evidence, doing their best, by innuendo and suggestion, to write down this new Eldorado, missed the essence and spirit of everything about them.

The human tide that every hour of the twenty-four ebbs and flows through the streets of Goldfield carries not mere wreckage or flotsam and jetsam. The salt of the earth is here as well as its scum. On the frontier, a man soon finds his level. No higher type of manhood can be found anywhere than these mine owners in a camp where it is said "a million dollars is the unit of the account—you hand it over, and it is not polite to ask for any change." To this type belongs Taylor, who sold the control of the Red Top and Jumbo mines to Wingfield and Nixon for $1,200,000. There was not a scrap of paper to bind the bargain. After the verbal option was given, and before it was exercised, the market quotation of the mines showed a paper profit of $2,000,000 to the purchasers. Did this apparent loss on the one side and gain on the other cause Taylor to swerve from his word, or, as they say in Nevada, fail to "come

Photo by Allen

ELECTION DAY IN GOLDFIELD

through?" He might easily have demanded and received half a million additional if he had chosen to enter the ranks of "welchers," a word which is well understood on the coast since the San Francisco fire—but Taylor was made of different stuff. It probably never occurred to him to "hold-up" or "shake-down" his old-time friends, who relied on his honor and good faith.

Myers, "the father of Goldfield," of Mohawk and Combination fame, is made of the same stuff. He passed his word to deliver 100,000 shares of Mohawk to Carstairs, a Philadelphia capitalist, for $400,000 in cash, providing the money was forthcoming in five days. Before the end of that time the market price of the stock showed a profit to the purchasers of nearly half a million dollars, and a few weeks later the share of the mine which he disposed of for $400,000 was worth, in open market, $2,000,000. His precious Mohawk, which he had held intact since he had located the mine, was not deposited in escrow with any bank or trust company. He had not received any payment on account, he merely passed his word to deliver it under certain conditions. These conditions being complied with, subsequent gain or loss on either side affected him not at all. He probably regretted that he had not asked and received a higher price—but he tendered the stock, when and as due, on demand. There are many more in Goldfield like Myers and Taylor—jealous of their word and proof against any temptation to break it.

Others there are in this mixed throng of money seekers of whom

Photo by Dickerson

THE MAIN STREET OF BEATTY

men say: "Look out for him! He has a shady past; he was indicted back East, and is out here under an *alias*, or masquerading under some new firm name. His capital consists of unlimited assurance and investors' lists, made up of thousands of names of men and women of small means who buy mining or other shares, led thereto by cunningly worded circulars, which he is an adept in preparing." What will you? These are the jackals and ghouls of every industry which catches the public eye. *Caveat emptor.* These men get only the leavings of the honest and law-abiding, and are soon found out and passed on to the next camp until the United States gets too hot for them, when they turn up in South America or some foreign country, using the same arts on new victims everywhere.

Two questions are asked outside where one meets those whom the fame of Goldfield draws like a magnet: How can I secure a share of its prosperity? and how long will it last? Judging the future by the past, there are as many new fortunes likely to be made in 1907 as were made in 1906. Unless all signs fail, we are entering upon a season of speculative excitement greater than the West has ever known. The stakes to be played for in the future will be larger, and th small investor and operator must look farther afield.

Goldfield is not only a great mining camp itself, but the contagion and intoxication of rapidly multiplying fortunes induce a kind of splendid optimism among the owners. This spirit manifests itself in excursions unto the remotest corners of the State, and no district is so inhospitable that the touring car of the Goldfield capitalist will not seek to explore it at the first rumor of a new strike.

Photo by Allen

SOME NATIVES OF NEVADA

Goldfield men have developed Fairview, Wonder, Searchlight, Round Mountain, Bullfrog, Manhattan, Rose Bud—as far north as the copper camps of Ely, and south as far as Greenwater and Death Valley.

Goldfield is no place for the lazy or incompetent. The man of constructive imagination who sees opportunity and pursues it without knowing discouragement or fatigue—to such a one this primitive life has much to offer. The society of men like himself, the mysteries of untrodden deserts, and sky-piercing mountains, the development of natural resources, almost virginal and fresh from the Creator's hands.

Beyond yonder horizon may lie another Tonopah or Goldfield with new rewards for the tireless Argonaut.

The world wonders at the reports of new strikes following upon one the heels of the other in this desert country. Why were these discoveries not made by the Mackays, Floods, O'Briens and their followers, who lifted Nevada so high before the world's gaze less than two-score years ago? This is the reason. Formerly the prospector, with his faithful burro, journeyed timidly from spring to spring, not daring to extend the area of his search beyond a circle a few leagues in extent. The great mineral zones of Southern Nevada were thus secure from from his excursions. With the advent of the railroad new stations with water supplies sprang up. Where formerly one lonely man wended his solitary way through cañon and valley, now scores of well-equipped prospecting parties bring back tales of new discoveries miles from their base of supplies. As soon

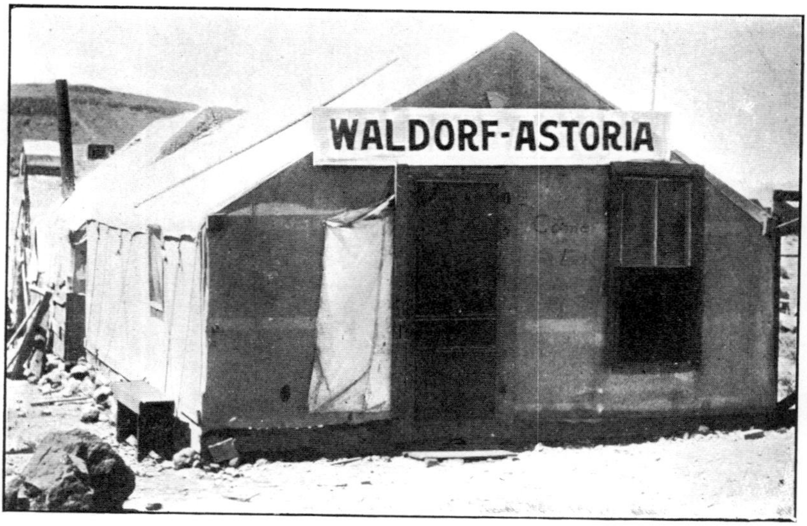

Copyright by A. Allen

AN EARLY TYPE OF HOTEL

as the news is verified, the capitalist is hurried thither in his automobile, tents rise as if by magic, water-pipes are laid from the nearest creek and the world learns that another mining camp has been found in the desert.

Under former conditions such rapid development would have been impossible. Railroad transportation has wrought this change, and as new lines are built and extended, other deposits of silver and gold and copper will be found from the development of which will spring another crop of millionaires.

Goldfield is well situated to become the center of this new domain. It holds the key to the southern country. There it seems destined to become another Denver. It is not lacking in natural advantages despite its isolation and present environment. Those who live in it the year around find no fault with the climate. It is neither too hot in

THE PROSPECTOR EN ROUTE

summer nor too cold in winter. Those who have taken great fortunes out of the ground are loyal and bent upon building a city both habitable and attractive.

If there was any chance that the new railways planning to reach Goldfield would by competition reduce freight rates, the city would take on a magical growth. The present rates and the high prices of labor handicap progress, and in the past men have been too busy picking money out of the ground to take any thought how they were housed and fed.

As to the permanence of the camp, let those who are doubtful read the reports of such well-known geologists and mining engineers as John W. Finch and Charles C. Moore, men who have lived there for a long enough period to pass an intelligent judgment.

Goldfield, Nev.

"STRUCK IT AT LAST"

By E. M. HAMILTON.

BEING out of health, I concluded to go camping; so fixed up a one-horse rig, took one of my sons with me and started out for the desert or any place that we might happen to go. I fetched up on the north side of Antelope Valley, about ninety miles north of Los Angeles. Being of an industrious disposition, I commenced to prospect. I prospected a little different from most prospectors. I saw a red looking hill on the north side of the valley. I liked the looks of it, so I took my horse and wagon and went in the gulches. They, being short and steep, were easy to prospect. I took a sack of dirt and called it No. 1, leaving a mark so that I could find the place in case there was anything in the dirt. I kept this up until I got the wagon loaded, then I drove three miles to a place where there was water and panned the dirt out. I found free gold in all the gulches. Then I went on the north side of the hill and done the same. I also found gold. I had learned to track Indians in the early times and read signs. I looked at that gold. It was rough and lay among broken stones, not gravel. I says to myself, it's a native of this place. Like the Irishman, I had it surrounded, so I was not long in finding the lead. I happened to find the lead in the barest place. There was a dike or large bunch of rock that assayed about $35.00 per ton. I thought there was gold enough in the dike to pay for a two-stamp mill, so I traded places and took money that belonged to the family and built the mill. When I came to break up that dike, if the gold had been put on with a whitewash brush it could not have been much thinner, so I was in debt, mill on hand and no pay-rock or ore. I did not give up. I kept on. My money was gone and so was my friends. Family helped me for awhile. They were short of means of support. My boys left me. Said there was nothing there. I worked all alone, sunk a shaft on the lead. The ore run about $2 per ton. I would carry the ore up on my back out of the shaft. Sometimes I was so weak I hardly could get a sack out of the shaft. I managed to get out about three-quarters of a ton. This made a load for the horse. I got out one load per day. I kept this up until I got out about ten tons at the mill, then I would take the horse and wagon and gather sage-brush to make fuel for the mill. Then I would mill it, get about $20 out of the ten tons and buy some more pork and beans and do so again. I mixed sage-brush with the hay to make it last. I kept this up for about two and one-half years. I had hopes of finally striking it. There is a rule among miners, if you have a color, follow it. I finally became worn out financially, physically and almost mentally. If I had got hurt, I might have laid there and perished, for there was no one near me. I was about to give up when the idea occurred to me, why, go down on a lower level, the lead runs diagonally across the hill. I went and looked around. I saw nothing that looked as though it had anything in it more than I saw a little clay sticking to the rock that looked as if it might be a foot-wall. I scratched down with my hands and took a handful of dirt where I had water. Found it must have had twenty-five cents in the horn. Went back, found it scattered all around. Just then, I would not have swapped places with Gov. Gage. I did not care so much for the money value, but I had triumphed over all difficulties. I sacked twenty-one tons of dirt, had to pick the roots out of it while sacking it. I sent it to the Selby Smelting Works. It brought me $4500 net in sixty days. I did not owe anyone a cent. Everything being so unusual, I placed myself in the exact position in which I was when I found the gold in the horn and had a kodak picture taken, which might be titled "From Poverty to Wealth" or "Struck It At Last." I have taken out some $200,000 and the mine is hardly commenced to be worked. I own the mine all alone. I have been offered large sums of money for it, but to give me a large sum of money at my age it would cause me lots of trouble, but where it is it causes me no trouble.

Rosamond, Cal.

The photograph to which Mr. Hamilton refers is reproduced as the frontispiece of this number of Out West. His account of the circumstances which led up to the "strike" is printed verbatim, partly because it is an interesting and truthful record of facts—partly as a set-off against the traditions of the discovery of great mines by a mule, a boot-heel, or a chance revolver-shot, some of which appear on other pages of this magazine. Chance is a factor in the prospector's life, but the grimmest kind of hard work is by far the more conspicuous factor.—*Editor.*

INDEX.

INDEX.

INDEX.

NEVADA GHOST TOWNS & MINING CAMPS, by S. W. Paher. Large 8½ x 11 format, 492 pages, 700 illustrations. About 668 ghost towns are described with directions on how to get to them. It contains more pictures and describes more localities than any other Nevada book. Nearly every page brings new information and unpublished photos of the towns, the mines, the people and early Nevada life. It is the best seller of any book on Nevada ever published and it won the national "Award of Merit" for history. $30.00.

LAS VEGAS, AS IT BEGAN, AS IT GREW, by S. W. Paher. Large 9 x 12 format, 181 pages, 210 duotone illustrations, maps, index, biblio. In the 19th century Las Vegas was first a Spanish Trail waterhole, then a religious mission and then a ranch. Though founded in 1905 as a railroad town, it developed into a gambling center during distinct boom periods. The reader is led every step along these transitions in both the text and photos. $20.00

HISTORY OF NEVADA, 1540-1888, by H. H. Bancroft. 332 pp, 6 x 9. This is a reprint of the 1888 history of the early explorers through Nevada through the beginning of statehood in 1864 up to the date of publication. Large chapters discuss the Comstock Lode, state government, Indian wars, transportation, etc. $25.00.

HISTORY OF UTAH, 1840-1886, by H. H. Bancroft. 808 pp, illus., maps. Utah became a territory in 1850, three years after the Mormons came to settle Salt Lake Valley. Here is a detailed history of the early growth and development of Utah. Of interest to Nevadans is material on Carson County, western Nevada. $25.00.

HISTORY OF NEVADA, 1881, Thompson & West, ed. by Myron F. Angel. 900 pp, 7 x 10. This reprint of the 1881 history has 35 chapters covering every important development in Nevada's formative years — mining development, Indian raids, raucous politics, mail and press, early exploration, social organizations, etc. Granddaddy of Nevada historical works, and a researcher's must. $60.00.

INDEX TO HISTORY OF NEVADA, 1881, Thompson & West, by Helen Poulton. 160 pp, 7 x 10. This index of more than 10,000 entries is a most valuable auxiliary tool to the above history. This second edition adds a clear map. $8.00.

NEVADA, AN ANNOTATED BIBLIOGRAPHY, by S. W. Paher. 585 pp, 7 x 10, illus. Here is a researcher's guide and description of more than 2,544 books relating to the history and development of Nevada. The useful 74-page index has 3,550 subjects; 33 other bibliographies refer the researcher to countless Nevada materials. Printed on high quality paper and buckram bound. $95.00.

Write for catalog of more than 100 ghost town and treasure books on the Southwestern states.

U.S. Geological Survey Mining Bulletins Reprint Series

These bulletins, issued early in this century, are much in demand as primary reference material for Nevada mining and water sources. All cover extensively mineralogy, topography, history and the various mines. All are 6 x 9 quality paperbacks with pages sewn together. $14.95 each.

MINES OF THE GOLDFIELD, BULLFROG AND OTHER SOUTHERN NEVADA DISTRICTS, by F. L. Ransome. 144 pp, maps, index. Color map 22 x 28 laid in. This 1907 Bulletin also covers Searchlight, Eldorado, Crescent and Gold Mountain. The last 42 pages reproduce from Out West magazine two contemporary illustrated articles, "The Nevada Bonanzas of Today" and "Goldfield and the Goldfielders." These capture the flavor and excitement of Nevada's mining era.

MINES OF THE SILVER PEAK RANGE, KAWICH RANGE AND OTHER SOUTHWESTERN NEVADA DISTRICTS, by S. H. Ball. 218 pp, maps, index. Color map 22 x 28 laid in. This 1907 Bulletin also covers Lida-Goldfield valley, Pahute Mesa, Cactus Range, Reveille Range, Belted Range, Tolicha Peak, the Amargosa Mountain system, Gold Mountain, Death Valley and the Panamint Range.

320 DESERT WATERING PLACES IN SOUTHEASTERN CALIFORNIA AND SOUTHWESTERN NEVADA, by W. C. Mendenhall. 104 pp, illus, index. Color map 14 x 20 laid in. This 1909 Water-Supply Paper covers the Death Valley Basin in Inyo County and the Salton Sink of Imperial County, and numerous places in between. There are descriptions of 323 springs, all numbered and keyed to the colored map. The bulk of the book concerns San Bernardino County; in Nevada the following counties: Esmeralda, Nye, and Clark.

MINES OF BATTLE MOUNTAIN, REESE RIVER, AURORA AND OTHER WESTERN NEVADA DISTRICTS, by J. M. Hill. 200 pp, maps, index. This 1915 Bulletin also covers California's Lassen and Modoc counties, Douglas County, the Peavine district, and in Mineral County the Hawthorne, Granite, Santa Fe, Silver Star, and Pine Grove in Lyon County.

MINES OF TUSCARORA, CORTEZ AND OTHER NORTHERN NEVADA DISTRICTS, by W. H. Emmons. 220 pp, maps, index. This 1910 Bulletin also covers Midas, Good Hope, Columbia, Edgemont, Mountain City, Bullion, Safford, Tenabo, Lander, and the Lewis districts. Included is F. L. Ransome's "Notes on Some Mining Districts in Humboldt County" [Pershing County] 1909. Districts included are: Seven Troughs, Rosebud, Red Butte, Humboldt Range, Pahute Range, Sonoma Range.

MINES OF CHERRY CREEK, BRISTOL AND OTHER EASTERN NEVADA DISTRICTS, by J. M. Hill. 214 pp, maps, index. This 1916 Bulletin also includes Gold Butte, the Ruby Range, Mud Springs, Spruce Mountain, Mizpah, Kinsley, Toano Range, Tecoma, Ravenswood, Atlanta, Patterson, Troy, Bald Mountain, Granite, Hunter, Ward, Aurum, Duck Creek, Taylor and Kern districts.

MINING DISTRICTS AND MINERAL RESOURCES OF NEVADA, by Francis C. Lincoln. 295 pp, maps, index. This compilation gives a summary of each mining district in Nevada through 1923. There are historical summaries and bibliographies of each mining area, all arranged by counties. Section Two describes Nevada's mineral resources by types. An essential tool for the mining researcher. $14.95.

PONDEROSA COUNTRY, by S. W. Paher. 48 pp, large 9 x 12 format, two-color illus, historic and modern. Color cover. Tours out of Reno guide the traveler to Virginia City, Carson City, Lake Tahoe, and ghost towns. Early Reno pictures included. Clothbound, $7.50.

CENTRAL ARIZONA GHOST TOWNS, by S. W. Paher. 48 pp, 9 x 12 map, 50 old-time photographs. Written in collaboration with Robert L. Spude, this book summarizes in text and pictures the mining camps between Phoenix and Prescott, including those in the Bradshaw Mountains and Wickenburg and Jerome areas. Maps and directions are a clear aid for ghosttowners. $3.95 paper; $7.50 cloth.

NORTHWESTERN ARIZONA GHOST TOWNS, by S. W. Paher. 32 pp, 9 x 12. Mining camps in the Las Vegas-Kingman area are described in a lively text with historic photographs. Oatman, El Dorado, Kingman, Searchlight, Chloride, and about 20 other ghost towns are included. $2.95.

COLORADO RIVER GHOST TOWNS, by S. W. Paher. 80 pp, full color cover, 50 old-time photographs, 20 pages of color, cloth (with dust jacket). $9.95.

CHLORIDE MINES AND MURALS, by Roy Purcell. 16 pp, 9 x 12. History of the silver camp of Chloride in northwestern Arizona is recalled by lifelong resident, with illustrations. Artist Purcell also interprets his Chloride murals. Color cover. $1.95.

TOMBSTONE, ARIZONA SILVER CAMP, by Bob Spude. 16 pp, 9 x 12. Author recounts the discovery, the boom days of the early 1880's (with illustrations), the mines, the town and its legendary violence. Color cover. $1.95.

DEATH VALLEY JUNCTION, THE STORY OF THE AMARGOSA OPERA HOUSE, by Marta Becket. Author tells her own story in 35 pictures (7 in color, including cover) of the development of an opera house complete with historic murals. $2.50.

DEATH VALLEY GHOST TOWNS, VOL. I, by S. W. Paher. 32 pp, 9 x 12, map, 50 old-time photos, color cover. Though Death Valley's mineral history is best known for its colorful eras of borax mining, there were two distinct periods of gold and silver mining also. The first one included Panamint and Calico, while the early 20th century boom produced Rhyolite, Greenwater, and others. About 35 mining camps are included. $2.95 paper.

DEATH VALLEY GHOST TOWNS, VOL. II, by S. W. Paher. 32 pp, 9 x 12. Mining camps of the Death Valley National Monument — Skidoo, Panamint City, and Old Stovepipe Wells — are joined by those immediately to the west, including Cerro Gordo, Darwin and Cartago. Essay on "The Prospector," the Tonopah & Tidewater Railroad, and desert driving hints. $2.95.

DEATH VALLEY GHOST TOWNS, by S. W. Paher. 48 pp, 9 x 12, map. The heart of the above two books combine to form this hardbound book. $7.50.

SOUTHWEST ARIZONA GHOST TOWNS, by S. W. Paher. 32 pp, 9 x 12. Within a day's drive of Yuma are important camps dating from the late 1850's: Gila City, LaPaz, Ehrenburg, and many more. There is information about the steamboats that plied the Colorado River upstream from Yuma. The colored cover etching of a placer mining camp; the back cover is an 1864 Bancroft map in color. $2.95.

SCOTTY'S CASTLE, DEATH VALLEY, by S. W. Paher. Large 9 x 12 format, 48 pp, heavily illustrated in color. Besides a history of Death Valley Scotty, there are many intricate pictures of the castle itself, with large captions to explain them. $3.75 paperback, with color cover.

THE CAMELS OF NEVADA, by Douglas McDonald. 32 pp, illus. In Nevada camel pack trains hauled salt, wood and even freight, also aiding early surveyors. But the beasts also brought problems. Modern camel races in Virginia City are recounted. $2.95 paperback.

THE BIG BONANZA, by Dan DeQuille (William Wright). 488 pages, illustrated, with intro. by Oscar Lewis. Indexed. Subtitled "An authentic account of the discovery, history, and working of the Comstock Lode," the Big Bonanza by DeQuille covers every phase of the epic rise of Virginia City, especially the special technology required to work the deep silver mines. $12.95.

GHOSTS OF THE GLORY TRAIL, by Nell Murbarger. 316 pages, illus. Indexed. Subtitled "Intimate glimpses into the past and present of 275 western ghosttowns," Ghosts of the Glory Trail is a fast-moving chronicle depicting the early-day mining stampedes. All Nevada counties are represented either in the 39 chapters on specific towns (such as Aurora, Rhyolite, Candelaria, Tuscarora, Delamar, etc.) or in the valuable ghost town directory containing 225 additional listings, some in eastern Calif. and western Utah. $12.95.

THE HISTORY OF NEVADA, by Sam P. Davis. 2 volumes. 1279 pages, illus. Reprint of the rare 1913 first edition. There are chapters on every phase of Nevada history including Indians, topography, the emigrants, mining, politics, journalism, education, religion, railroads, medicine, banking, agriculture, fraternal societies, drama, reclamation, divorce law, and a chapter on every Nevada county. $60.00.

NEVADA POST OFFICES, AN ILLUSTRATED HISTORY, by James Gamett and S. W. Paher. 160 pp, 7 x 10, illus. About 760 Nevada towns had post offices and here is a detailed list of them. There are chapters of the Pony Express, Wells-Fargo, and how to collect postal materials. 110 postal illustrations are in this research tool. Hardcover with color dust jacket. $30.00.

NEVADA GHOST TOWN TRAILS, by Mickey Broman. 80 pp, illus., maps. Quick reference and guide to 120 ghost towns. $2.95. Also available in same format is **CALIFORNIA GHOST TOWN TRAILS.**

GEM TRAILS IN CALIFORNIA, by Mickey Broman, 84 pp, illus., maps. Good maps lead traveler to numerous gem stone collecting areas. $2.95.

GOLD FEVER AND THE ART OF PANNING AND SLUICING, by Lois DeLorenzo. 80 pp, illus. A how-to book for the amateur. $3.50.

WHERE TO FIND GOLD AND GEMS IN NEVADA, by James Klein. 112 pp, maps, illus. Gold and gem sites in every county are included. $5.95.

THE WEEKEND GOLD MINER, by A. H. Ryan. 64 pp. A simple, easy-to-follow handbook. $1.95.

YOSEMITE TRAILS, by Lew and Ginny Clark. 144 pp, illus., maps, index. Complete guide to this National park. $5.95.

HIGH MOUNTAINS AND DEEP VALLEYS, by Lew and Ginny Clark. 191 pp, b&w and color illus., maps, index. Coverage is from Virginia City southward to all of eastern California (including Death Valley), and Tonopah-Goldfield-Beatty, Nevada. $6.95.

JOHN MUIR TRAIL COUNTRY, by Lew and Ginny Clark. 176 pp, illus., maps, index. A knowledgeable guide of the authors' beloved Sierra. $5.95.

COMSTOCK MINING AND MINERS, by Eliot Lord. 578 pp, illus., maps. Reprint of valuable 1883 mining report. $20.00.

RAILROADS OF NEVADA AND EASTERN CALIFORNIA, [Vol. 1, Northern Nevada; Vol. 2, Southern Nevada and Death Valley], by David F. Myrick. 469 and 492 pp, illus., maps, index. Valuable economic and railroad history with much politics and mining included. $30.00 each.

LOST LEGENDS OF THE SILVER STATE, by Gerald Higgs. Short stories on various Nevada topics. $8.95.

FORT CHURCHILL: NEVADA'S MILITARY OUTPOST OF THE 1860's, ed. by S. W. Paher. 48 pp, 6 x 9. Illus. In the 1860's the soldiers at the Fort protected westbound emigrants and isolated Nevada settlements from Indian raids, and assisted Nevada's governors after prison breaks, miner's strikes, and lynchings. The color cover features the ruins of adobe buildings while on the back is a rare 1868 map in color. $2.95.

THE LEGEND OF JULIA BULETTE AND THE RED LIGHT LADIES OF NEVADA, by Douglas McDonald. 32 pp, illus., map. The author has put together interesting photographs and the best written historical sketch to date of Virginia City's famed prostitute, murdered in 1868. An overview of Nevada prostitution occupies the last part of the book. $2.95.

NEVADA LOST MINES AND BURIED TREASURE, by Douglas McDonald. 128 pp, 6 x 9. Illus. Color cover. Legends of lost mines in Nevada date from the Gold Rush of 1849 when westbound emigrants discovered silver in the desolate Black Rock Desert. In all, the author recounts 74 of these stories which also include tales of buried coins, bullion bars, stolen bank money, etc. Seven pages of two-color maps show general localities of the forgotten treasures. $5.95.

CALLVILLE, HEAD OF NAVIGATIO — ARIZONA TERRITORY, ed. b S. W. Paher. 40 pp, 9 x 12. Maps, illus trated. In the 1860's Callville was a shippin point for Utah and Nevada settlement Here is the story of steamboat activity o the Colorado River, and area mines and In dians. Text and 24 color pictures show Co orado River lore and man's intrusion. $4.9

Treasure Hunters Guide Series

by Theron Fox. All are 24 pp, 6 x 9, illus with old photos and 1881 map folded in pocket. These handy guides are to be use along with a modern road map. $1.95 each

ARIZONA TREASURE HUNTER GHOST TOWN GUIDE. Early maps o 1867 and 1881 show 1200 place names — mining camps, mountains, rivers an creeks, railroads, abandoned roads, forts Indian reserv., mines, canyons and passes

EASTERN CALIFORNIA TREASUR HUNTERS GHOST TOWN GUIDE Early maps of 1860 and 1881 show 70 place names — mining camps, abandone roads, springs, mountains, rivers, lakes water holes, etc. in Alpine southward t San Diego County.

NEVADA TREASURE HUNTER GHOST TOWN GUIDE. Early maps o 1867 and 1881 show 800 place names — mining camps, abandoned roads, springs mountains, rivers, lakes, water holes, an even a camel trail.

UTAH TREASURE HUNTERS GHOS TOWN GUIDE. Early maps of 1867 an 1881 show 900 place names — minin camps, old roads, mountains, rivers, lakes water holes, and springs.

Nevada Publications
Box 15444
Las Vegas, Nev. 89114